Urban Agriculture

The *Urban Agriculture* Book Series at Springer is for researchers, professionals, policy-makers and practitioners working on agriculture in and near urban areas. Urban agriculture (UA) can serve as a multifunctional resource for resilient food systems and socio-culturally, economically and ecologically sustainable cities.

For the Book Series Editors, the main objective of this series is to mobilize and enhance capacities to share UA experiences and research results, compare methodologies and tools, identify technological obstacles, and adapt solutions. By diffusing this knowledge, the aim is to contribute to building the capacity of policy-makers, professionals and practitioners in governments, international agencies, civil society, the private sector as well as academia, to effectively incorporate UA in their field of interests. It is also to constitute a global research community to debate the lessons from UA initiatives, to compare approaches, and to supply tools for aiding in the conception and evaluation of various strategies of UA development.

The concerned scientific field of this series is large because UA combines agricultural issues with those related to city management and development. Thus this interdisciplinary Book Series brings together environmental sciences, agronomy, urban and regional planning, architecture, landscape design, economics, social sciences, soil sciences, public health and nutrition, recognizing UA's contribution to meeting society's basic needs, feeding people, structuring the cities while shaping their development. All these scientific fields are of interest for this Book Series. Books in this Series will analyze UA research and actions; program implementation, urban policies, technological innovations, social and economic development, management of resources (soil/land, water, wastes…) for or by urban agriculture, are all pertinent here.

This Book Series includes a mix of edited, coauthored, and single-authored books. These books could be based on research programs, conference papers, or other collective efforts, as well as completed theses or entirely new manuscripts.

Hiroyuki Tsunashima

Farming and Social Inclusion in Japan

From Welfare-to-Work to Delight

Springer

Hiroyuki Tsunashima
Osaka Metropolitan University
Urban Resilience Research Center
Osaka, Japan

ISSN 2197-1730 ISSN 2197-1749 (electronic)
Urban Agriculture
ISBN 978-3-032-01597-6 ISBN 978-3-032-01598-3 (eBook)
https://doi.org/10.1007/978-3-032-01598-3

This work was supported by Japan Society for the Promotion of Science

This Springer imprint is published by the registered company Springer Nature Switzerland AG
The registered company address is: Gewerbestrasse 11, 6330 Cham, Switzerland

Acknowledgments

This work is dedicated to the memory of Nagase Shigeru and Hisatsune Yasunori, who, due to their homelessness, resided in Nishinari Park, Osaka City. Their invaluable contributions were instrumental in initiating the action research project that serves as the foundation for this volume.

In addition, the following interviewees and collaborators, listed alphabetically, have helped, as it were, co-author this text: Arai Toshimasa, Bōshita Akenobu, Bōshita Kayoko, Fujisaki Yoshihiro, Fujita Minoru, Fukushima Terumi, Hanai Yūta, Hashimoto Takuya, Hayashi Kazuma, Higa Hiroshi, Higuhi Shūichi, Hosokawa Masayo, Hōzawa Nanigashi, Igarashi Kēta, Īda Sayuri, Inagaki Hiroyuki, Inoue Kyōichi, Inoue Noboru, Kasagi Yasutaka, Kasahara Yutaka, the late Kashima Noritake, Kataoka Akiyoshi, Katsu sōshi, Kobayashi Makoto, Kohama Yukitaka, Kotegawa Nozomi, Matsumoto Nanigashi, Matsunaga Masaru, Mukae Ryōji, Nakagiri Kōsuke, Naobayashi Yū, Narukawa Teruaki, Nishimori Masaaki, NĪhara Ryō, Nohechi Akane, Nomoto Toshimitsu, Ojima Kiyoko, Okada Fukuji, Ōshima Haruo, Sakaguchi Natsuki, Sakumoto Hiromi, Satou Yoshimasa, Seta Hiroki, Shimizu Takurō, Sōda Nobutsugu, Sugi Teruaki, Sugiue Yōhei, Suzuki Kazuhiro, Takaoka Mie, Takemoto Kēta, Tamaki Nanigashi, Tanaka Masanori, Tonohata Satoshi, Tsuji Naoki, Uemura Tsutomu, Uemura Yōichi, Wada Akio, Waku Kanamu, Yamaguchi Shinji, Yamamoto Kazuhiro, Yamamoto Kenji, Yamamoto Kenji, Yamamoto Tadakazu, Yamashita Kōhē, Yoshida Sakiko, and Yūkawa Masashi.

I would also like to express my gratitude to the following organizations: the Organization to Support the Homeless in Kamagasaki, an approved nonprofit organization; Cooperative Institution for Vocational Education and Training of Osaka Limited Liability Partnership aka "A Dash Work Creation Studio," the former Next Stage Osaka Limited Liability Partnership; Nishinari Ward Social Welfare Council; Osaka City's Association of Disability Welfare and Sports Social Welfare Corporation; RE Work Together General Incorporated Association; and Specified Nonprofit Organization Foro.

This work was supported by the following funders: Japan Health Foundation (research grants in fiscal year 2017); the Toyota Foundation, Japan (D18-R-0360); Japan Society for the Promotion of Science (KAKENHI Grant Nos. 23730519,

25705015, 19K06297); and the Univers Foundation (18-03-173). Additionally, I received assistance from the clerical staff members of the Urban Resilience Research Center, Osaka Metropolitan University, namely Hatoko Naomi, Hori Takaaki, Igarashi Shizuko, Kim Mina, Kojima Naomi, and Takami Asuka.

I am also grateful to the staff members at Springer and the editors of the Urban Agriculture Series, whose continued support has been invaluable. In particular, I would like to express my sincere gratitude to Prof. Christine Aubry for her insightful and encouraging comments throughout the development of this manuscript. I lack the words to fully express my gratitude to these compassionate colleagues. Furthermore, I thank Editage for English language editing.

Finally, my two children have always been a source of thought-provoking inspiration. They continually challenge me to question what it means to nurture life. I owe this remarkable natural phenomenon to their mother, cousins, friends, former childcare workers, and later teachers, who have been integral to their growth.

Competing Interests The author has no competing interests to declare that are relevant to the content of this manuscript.

Contents

Chapter 1
Introduction: A Theoretical Challenge in Justifying the "Partnership Between Agriculture and Social Welfare"

Abstract This volume begins by introducing a newly coined term in Japan, which translates to "partnership between agriculture and social welfare." While it might be seen as analogous to terms like "care farming" or "social farming" used in the West, it differs in its explicit reference to "social welfare." The rationale for collaboration between "social welfare" and "agriculture" remains unclear, and whether this movement can be integrated into social welfare systems is still uncertain. The primary aim of this volume is to provide theoretical foundations for this partnership from a distinct perspective. This introductory chapter outlines the key themes to be explored throughout the text.

Keywords Care/social farming · Employment · Ministry of Agriculture, Forestry and Fisheries (MAFF) · Ministry of Health, Labor and Welfare (MHLW) · Well-being

1.1 How the Concept of "Partnership Between Agriculture and Social Welfare" Is Currently Communicated

More than a decade has passed since the term "partnership between agriculture and social welfare" was introduced, and the movement has recently gained momentum in Japan. Notably, the Ministry of Agriculture, Forestry and Fisheries (MAFF) and the Ministry of Health, Labour and Welfare (MHLW) have formed an alliance to promote the initiative, based on the idea that the shorthanded agricultural sector provides job opportunities for those excluded from the labor market (e.g., Hamada 2018; Yoshida 2019). What is particularly striking is that the concept has gained such popularity that even ordinary enterprises now propose it as a strategy for managing plant factories in terms of profitability (e.g., BASF Japan Inc. 2020; SMART AGRI's Editors 2019a). Additionally, MAFF and MHLW jointly held "the first symposium on the agriculture-welfare collaboration that target the needy in Japan" in February 2021. However, MAFF (2020) acknowledges that no precise definition

H. Tsunashima, *Farming and Social Inclusion in Japan*, Urban Agriculture,
https://doi.org/10.1007/978-3-032-01598-3_1

of the term has been established. This ambiguity becomes clearer when we analyze the structure of the term itself.

The concept consists of three parts: "agriculture," "social welfare," and "collaboration." The first two are the agents of the latter; that is, they do not collaborate directly in a literal sense. "Agriculture" refers to the agricultural sector, including individual farmers, agricultural corporations, and their relevant administrative bodies. The second term, "social welfare," may also appear to directly correspond to the social welfare sector. Indeed, MHLW and social welfare organizations are involved in the movement. However, it is crucial to define technical terms independently of the popular usage or associations they may have.

The term "social welfare" does not have a self-evident meaning, especially since the existing systems lack a comprehensive theory to justify their structure. One of the simplest definitions available is "the aid provided by the government or private organizations to help the socially vulnerable" (The Third Edition of Daijirin, one of the most renowned Japanese dictionaries). However, this definition is somewhat tautological, as "the socially vulnerable" refers to individuals who need services. While traditionally, consumer goods and services were provided, today, job opportunities are increasingly offered as part of the Welfare-to-Work initiative. This shift followed a revision of the Disabled Persons Fundamental Law in 2004, along with a supplementary resolution mandating the government to create more local employment opportunities for disabled individuals. Subsequently, the 5-Year Plan for Priority Measures aimed to expand employment for disabled people in agricultural corporations (National Institute for Rural Engineering 2008). This movement undoubtedly contributes to the goals of the social welfare system. However, if "agriculture" is positioned as the counterpart to "social welfare" in this movement, several theoretical issues arise.

First, like the broader institution of social welfare, the movement will eventually be questioned about the criteria for selecting its target population. This issue is inextricably linked to the question of what social welfare entails. At present, it is unclear who should be the target of these initiatives. This ambiguity is reflected in the frequent use of the phrase "etc." by various organizations when explaining the target groups for the movement on their websites. Consequently, it is unclear whether the movement and the social welfare system share the same target population, both currently and ideally. More specifically, the question arises as to whether the movement should establish a unique definition of its beneficiaries, independent of how the social welfare system has historically defined them.

Second, the movement will be questioned on how "social welfare" assists its target population in collaboration with "agriculture." More specifically, how can the movement be situated within the social welfare system? In principle, "the current social welfare systems try to promote the welfare of targeted individuals and families, in terms of medical care, housing, education, and recreation, through boosting up their income to a certain extent" (Britannica Japan). However, this account does not clarify why "agriculture" should be considered a counterpart to "social welfare," as both primarily offer job and income-boosting opportunities. In fact, an ongoing issue remains unresolved: Employment Support Offices for the Disabled, for example, face significant challenges in increasing remuneration for participation to even reasonable levels, a problem that has persisted since the inception of the movement. The only area of agreement between the agricultural and social welfare sectors is that the former ensures adequate labor for farms, while the latter provides employment opportunities for the unemployed. This relationship appears to be a temporary and incidental interdependence, which could easily be disrupted if a more profitable industry were to experience a labor shortage.

Third, social welfare organizations often highlight the benefits participants can gain from agricultural activities from a different perspective. For example, some websites state that "being in contact with nature and growing crops enables participants to undergo effective mental rehabilitation" (Daiwa Lease Inc. 2020) or "a variety of people may have opportunities to take part in local communities as a member" (SMART AGRI's Editors 2019b; Pal System Consumers' Co-operative Union 2019). These benefits are typically grounded in the idea that agricultural activities promote a sense of well-being or happiness, consistent with the concept of "horticulture for well-being" advocated in Japan since earlier (cf. Matsuo 1998, 2005). However, these outcomes do not always translate into job opportunities or increased income for participants. Thus, such forms of well-being fall outside the primary scope of the existing social welfare systems. For the movement to justify these benefits within the framework of social welfare, it would need to align with the institution's broader goals. Currently, the movement seems to present these advantages haphazardly, without sufficient theoretical or empirical support.

1.2 Another Possible Reason for the Partnership

It is naive to assume that the social welfare sector can collaborate with agriculture merely in the name of "social welfare." To explore another possibility, it is necessary to first examine what is meant by "social welfare." There has been ongoing debate about the essence of social welfare. Some critics argue that, when social workers attempt to address the issues faced by vulnerable populations, they often rely on technical means that fail to tackle the root causes of these problems. As a result, it is nearly impossible for the social welfare sector to resolve these issues independently. This represents an intrinsic limitation of social welfare systems. To overcome this, social workers would need to engage in broader social movements, obtaining close cooperation from labor movements (cf. Kimura 2011, 2015). The idea that the root cause of these issues lies in capitalist production relations is supported by Suezaki (2018), who argues that the purpose of social welfare systems should be to counteract the social problems stemming from structural failures unique to capitalism. Both Suezaki and Kimura agree that social welfare systems cannot function effectively without the support of social movements to challenge the systemic flaws and resist opposing social forces. If this is the case, social welfare must target individuals whom existing systems often overlook. Paradoxically, the true social welfare beneficiaries are those who are rarely reached by institutionalized social welfare systems.

At this point, "social welfare" in the partnership context begins to also imply social movements. For any social movement to succeed, it must first identify the problems it seeks to address. However, the partnership movement has not yet clearly defined the specific social issues the social welfare sector should focus on. In line with the discussions of social welfare cited above, an agronomist has noted:

> The small agriculture in mountainous areas will remain undervalued and cannot survive in the future unless it approaches grand questions that face human beings, such as how to treat the illness that has originated from capitalism, and how to regain the humanity lost hand in hand with permeation of populism. ... Without proposing countermeasures against those problems from the standpoint which only the agricultural sector can adopt, the small agriculture will remain deemed as a subsidized industry or underrated as a burden to the industrialized society (Tsuno 1995).

Here, "small agriculture" refers to farming with lower labor productivity and, consequently, reduced international competitiveness. The tendency for small agriculture to be "underrated as a burden to the industrialized society" mirrors the situation faced by the social welfare sector. Importantly, this passage also suggests that small agriculture could help address problems stemming from capitalism and populism, though it does not specify how these ideas might be practically implemented.

This argument highlights another possible reason for the solidarity between small agriculture and social welfare: a shared structural commonality, rather than merely coincidental overlapping interests related to workforce demand and supply. Structural commonality does not inherently imply advantage, and this proposal may seem more idealistic than pragmatic, as though these two practices are destined to rely on each other in desperation. However, this potential solidarity is worth noting, as it distinguishes small agriculture from other industries. Genuine social welfare involves caring for those whose voices often go unheard, while the agricultural sector, in its smaller form, focuses on the care of non-human life, rather than the industrial production of commodities. This parallelism provides insight into another potential foundation for the partnership.

1.3 Another Question to Approach

We can begin by recognizing the intrinsic value of the partnership—this volume illustrates this through a concrete case, one I have presented for several years prior to the coining of the new catchphrase. The story dates back to the late 1990s when major cities in Japan experienced an unprecedented rise in homelessness. At the time, the government attributed this phenomenon to a temporary economic issue—mass unemployment resulting from the recession that began in 1991. As evidence, the Law for the Support of Self-Reliance by the Homeless, enacted in 2002, was initially set to be valid for only 10 years.

However, from a historical perspective, various policies had foreshadowed this development. The most recent was a policy favoring the casualization of employment, which essentially legalized pre-modern practices in the day labor market. Despite being prohibited by the Employment Security Law of 1947, the central government had long overlooked these practices. During the rapid economic growth of the 1960s, this labor market absorbed unemployment caused by the policy shift from coal to petroleum as the primary energy source (Fujino 2017; Shimazaki 2017). It also absorbed former farmers displaced by agricultural modernization, prompted by the Agricultural Basic Law of 1961 (Tōi 1963; Teruoka 2003). Moreover, the public assistance system, designed as a last resort in social welfare, failed to support casual laborers at risk of homelessness. This situation persisted until the 2000s when courts gradually required municipal governments to implement the system in strict adherence to the Livelihood Protection Law. In essence, the social welfare system excluded those unemployed workers deemed physically able to work.

Meanwhile, the agricultural sector faced a workforce shortage, as evidenced by the growing amount of uncultivated farmland, particularly since the 1990s (MAFF 2002). Involved in initiatives supporting homeless individuals in Osaka, Japan's largest city with the nation's largest day labor market, I came to believe that the urban-rural imbalance was a key factor linking these two phenomena. On both sides, I observed a social disdain for manual labor, often described "unskilled," as Tsuno (1995) had previously pointed out.

A local day laborers' union strongly advocated for alternative employment opportunities to address the rising unemployment. Union representatives demanded, "Just give us jobs back; we need nothing else!" They later reaffirmed that welfare benefits alone could not resolve the issue. The municipal government had already implemented a job provision scheme targeting homeless individuals and day laborers, in principle, aged 55 and older, offering positions in municipal facility maintenance, such as cleaning and mowing. Although the scheme was one of the positive outcomes of the union movement following the economic bubble burst of 1991, it was clear that the scheme was inadequate in terms of both quantity and, in my view, quality. I then conceived the idea of creating alternative agricultural jobs, hoping to simultaneously address the imbalanced urban-rural relationship.

Why did the union prioritize work over relief? A spokesperson explained that once former laborers became welfare recipients, they often found themselves with more time and energy than they knew how to use. They needed to regain something they had primarily obtained through work—whether that was personal connections with colleagues or a sense of belonging and contribution to society. In essence, they sought meaningfulness, a need that bare welfare benefits could not fulfill.

This emphasis on work aligns with the European Union's concept of "social inclusion," which recognizes employment not only as a source of income but also as a means of fostering social participation and personal development. Advocates of this concept argue that it is compatible with the Welfare-to-Work approach, as various Member States have implemented "activation" measures to eliminate disincentives to jobs and training for welfare beneficiaries (European Commission 2004).

However, some scholars argue that such "activation" may negatively impact social inclusion, especially if it narrowly focuses on enhancing the employability of job seekers (e.g., Perkins 2010; Higuchi 2014). Girardi et al. (2019) further argue that the creation of "public works" for social inclusion may perpetuate the stigma of being treated as irregular workers who receive social assistance. Ultimately, van Gerven et al. (2024) point out that effective "activation" governance requires cooperation among different stakeholders. This dynamic suggests that job creation initiatives, which inherently require collaboration between social welfare and other industrial sectors, may inevitably resemble "activation" projects. Indeed, the municipal creation of "public works" for day laborers in Osaka could be viewed as such, but it was, from the start, a product of union-driven efforts to preserve the workers' active status. Thus, the term "activation," derived from the transitive verb "activate," is not appropriate for describing their cordial aspirations. My proposal for supplementary agricultural jobs aimed to align with the workers' goals, not to "activate" them through external incentives.

In 2011, I began an action research project aimed at recovering uncultivated farmland with the involvement of homeless elderly laborers and young trainees receiving employment support. Preliminary results suggested that farmwork could offer an alternative to conventional employment assistance, which typically focuses on individual social adaptation. However, providing participants with opportunities to grow crops while paying them wages proved challenging. It may seem counter-intuitive, but I postulated that this question—whether the act of growing crops serves as an appropriate alternative job creation initiative—was central to the project. Additional questions emerged regarding whether, why, and how this initiative could address the urban-rural imbalance and contribute to social welfare. These questions remain unanswered in the current discourse on the partnership movement in Japan. At the same time, I realized that from an international perspective, this partnership may resemble intersectoral collaborations aimed at "activation" in European countries.

To explore this, we must look to overseas models such as care/social farming and horticultural therapy. Sempik et al. (2005) have already questioned why horticulture or gardening is used for social inclusion, providing insightful explanations, which have been expanded upon by subsequent studies (e.g., Gonzalez et al. 2011). However, several issues remain if we are to theoretically justify these activities as social welfare initiatives, as defined earlier. Past studies have primarily focused on the mental and physical well-being of vulnerable populations, without addressing how these activities contribute to social welfare. They tend to generalize human experiences, assuming a homogeneous society, while neglecting the fact that social welfare must address issues arising between groups with different socioeconomic positions. Without a clear definition of which social problems horticultural or gardening activities aim to address, it is impossible to assess how they contribute to social welfare. Ultimately, past research does not fully answer the questions raised during the action research project. Rather than seeking lessons from these studies, the current project may contribute to the fields of horticultural therapy and care/social farming.

The term "partnership between agriculture and social welfare," though challenging, provides an opportunity to reconsider how these two sectors should be combined. The essence of social welfare, as it relates to this partnership, emerges as a crucial conceptual tool for analyzing even international models like care/social farming, where different sectors collaborate for various motives (Hassink et al. 2016). This concept may prove valuable for practitioners and researchers as they share their experiences globally. The question "Why agriculture for social welfare, not vice versa?" is thus worth exploring in an international context.

1.4 Overview of This Volume

This volume primarily presents findings from the aforementioned action research project, with the methodology discussed in detail in the next chapter. Early achievements are explained through interviews with participants and my own observations. Additionally, an analysis of various cases within Japan—each differing in its degree of alignment with the partnership model—provides further insight into the project's key features. Consequently, early chapters may appear to place undue emphasis on the work of Japanese researchers. This focus reflects a strategic approach to integrating these partnership actions within Japan into a cohesive framework before contextualizing them internationally.

The findings from this project will not only critically review existing social inclusion schemes but also highlight the differing views between other practitioners and myself regarding the link between agriculture and social welfare. I aim to incorporate diverse perspectives while responding to criticisms of my own. Following this, the discussion will shift to how the action research project has evolved in response to these findings. The scope will then expand to an international context before concluding.

The organization of this volume is as follows: The next two chapters primarily review the action research project. Chapter 2 discusses the participants' perceptions of the value of working in a farming environment. Interview results suggest that farmwork differs significantly from urban occupations, with the key distinction being the natural laws governing sociocultural norms on the farm. This chapter ultimately highlights the dual benefits of introducing farmwork into employment assistance programs for young trainees. However, since most participants were urban dwellers, their interview responses did not reflect a full understanding of farming, a gap that I should have addressed more effectively. Chapter 3 provides an overview of the instructions I gave participants to maintain the farm at a certain production level. The subsequent analysis suggests that learning to grow crops autonomously requires changes in the dynamics between participants, their director, and the crops—a shift that proved difficult to facilitate during the job creation initiative.

To explore this issue, Chap. 4 examines several domestic progressive cases. For each case, I interviewed the director or managerial staff member overseeing client instruction, with a few exceptions where participants worked entirely autonomously. The results reveal that I am the only one particularly focused on autonomy, prompting me to clarify my reasoning. The chapter then explores the role of social welfare organizations in supporting clients in crop cultivation. My position is consistently maintained in Chap. 5. Drawing on insights from cognitive science and the philosophy of mind, particularly the "frame problem," the first half of this chapter addresses the challenges humans face when observing crop growth. The second half discusses later developments in the action research project, which has shifted its focus from financial benefits to providing participants with the opportunity to find joy in observing crop growth and sharing the harvest.

Chapter 6 argues that the findings offer valuable insights for international research communities, particularly in light of the growing literature on care/social farming, a model primarily practiced in Europe. The role of social welfare in such activities remains underexplored. Recent systematic reviews propose logic models for care farming tailored to different service user groups. However, the findings from this study suggest that these models could be grounded in fewer, more universally applicable principles related to crop cultivation. This approach would ensure that the role of care/social farming in social reform receives the attention it deserves. Chapter 7 examines future educational challenges related to the partnership between agriculture and social welfare. To date, agriculture and social welfare have been taught separately. However, as the partnership concept gains traction in Japan, cross-referencing between the two fields is likely to increase. Drawing from my experience as a university lecturer, I will present cases for consideration. I will also explore how to engage the broader public, not just the socially vulnerable, before concluding with a broader reflection on the concept of "social inclusion," which, in this context, takes on a reversed meaning.

The logical development in this volume differs from prior publications, as illustrated in Fig. 1.1. Mainstream discourses on the partnership in Japan, such as those

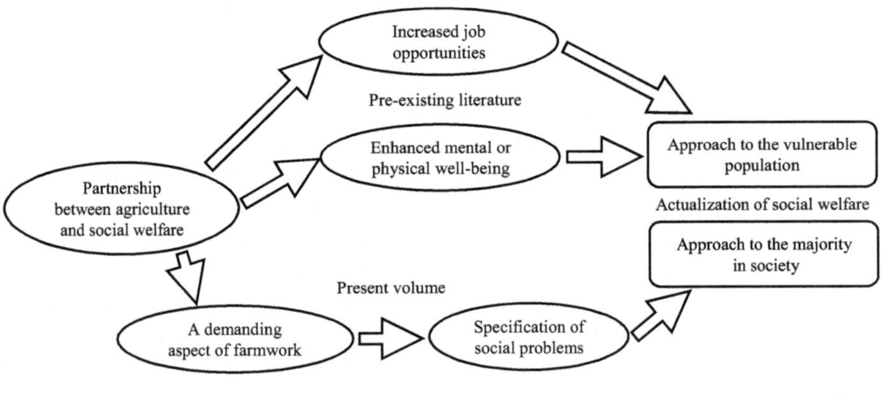

Fig. 1.1 Logical development in this volume in comparison with that in preceding publications. Arrows indicate a cause-and effect-relationship between the two connected occurrences. Each initial and final point represent the cause and the effect, respectively

led by MAFF and MHLW, emphasize economic benefits for both agriculture and social welfare sectors, including job creation and access to an inexpensive labor force. Some of these discourses also claim that such job opportunities improve the mental and physical well-being of workers, a position shared by advocates of care/social farming and horticultural therapy, particularly in the West. In contrast, this volume argues that the partnership should specifically address societal issues. The demanding nature of farmwork for urban novice horticulturists mirrors the exploitative aspects of the existing labor market and how workers adapt to these conditions. Recognizing this is essential for realizing true social welfare.

References[1]

***BASF Japan Inc. (2020) What is plant factory? Comments on its advantages and disadvantages. https://minorasu.basf.co.jp/80045. Accessed 20 Jan 2023

***Daiwa Lease Inc. (2020) Toward local community for coexistence through combination of agriculture and welfare: what is the partnership between agriculture and social welfare? https://www.daiwalease.co.jp/column/col_1163.html. Accessed 20 Jan 2023

European Commission: Directorate-General for Employment and Social Affairs (2004) Joint report on social inclusion. Office for Official Publications of the European Communities, Luxembourg

***Fujino Y (2017) Enforcement of the temporary law concerning coal mining industry rationalization measures [Sekitan kōgyō gōrika rinnji sotihō no sēritsu]. In: Annual report of Keiwa Liberal Arts Research Institute, vol 15, pp 69–106

Girardi S, Pulignano V, Maas R (2019) Activated and included? The social inclusion of social assistance beneficiaries engaged in "public works". Int J Sociol Soc Policy 39(9/10):738–751

Gonzalez MT, Hartig T, Patil G, Martinsen EW (2011) A prospective study of existential issues in therapeutic horticulture for clinical depression. Ment Health Nurs 32:73–81

***Hamada K (2018) Welfare agriculture in present situation and its future tasks and prospects: toward a wider horizon of agriculture, a new form of the agriculture-welfare partnership [Shakaihukusi no genjō, kadai, tembō: Nōhukurenkē no atarashī katachi, "nōsēgyō" e]. Agric Hortic 93(9):793–798

Hassink J, Hulsink W, Grin J (2016) Entrepreneurship in agriculture and healthcare: different entry strategies of care farmers. J Rural Stud 43:27–39

Higuchi A (2014) The mechanisms of social exclusion in modern society: the dilemma of active labor market policy. Int J Jpn Sociol 23:110–124

***Kimura A (2011) Social policy and "social security and welfare": target issues and system structure [Shakaisēsaku to "shakaihoshō, shakaifukushi": taishōkadai to sēdotaikē]. Gakubunnsha, Tokyo

***Kimura A (2015) Basic issues of theory about social welfare [Shakaifukushi no kihonmondai]. Gakubunnsha, Tokyo

***MAFF (2002) How abandoned farmland matters. https://www.maff.go.jp/j/study/nouti_seisaku/03/pdf/ref_data2.pd

***MAFF (2020) A textbook for training in technical support in collaboration of agriculture and social welfare [Nōfukurenkē gijutsu shiensha ikusē kenshū tekisuto] Ver.1

[1] *: in Japanese with an English title and summary, **: in Japanese with an English title, ***: in Japanese.

Matsuo E (1998) What is "horticulture wellbeing" in relation to "horticulture therapy"? In: Burchett MD, Tarran J, Wood R (eds) Towards a new millennium in people-plant relationships. University of Technology, Sydney Printing Services, Sydney, pp 174–180

***Matsuo E (2005) An exhortation to horticulture: environment, education, welfare and community development [Shakaiengēgaku no susume]. Rural Culture Association Japan, Tokyo

***National Institute for Rural Engineering (2008) A guide to employment of people with disabilities in the agricultural sector: case examples [Nōgyōbun-ya ni okeru shōgaishashūrō no tebiki: sagyōjirēhen]

***Pal System Consumers' Co-operative Union (2019) Fair local community materialized through the "partnership between agriculture and social welfare": the disabled, those who socially withdraw and the elderly are all community members. https://kokocara.pal-system.co.jp/2019/05/27/agriculture-and-welfare-collaboration-daiki/. Accessed 20 Jan 2023

Perkins D (2010) Activation and social inclusion: challenges and possibilities. Aust J Soc Issues 45(2):267–287

Sempik J, Aldridge J, Becker S (2005) Health, well-being and social inclusion: therapeutic horticulture in the UK. Policy Press, Bristol

***Shimazaki N (2017) Inter-regional job-seekers movement in the earliest stage of countermeasures for massive unemployment of former coal miners against: an analysis grounded on individual slip data of "reports on former coal miners' employment obtainments" [Tankōrishokushataisakushoki ni okeru rōdōsha no kōikishokugyō idō: "tankōrishokushashūshokutsūhō" kohyō dēta ni yoru bun-seki]. In: Research on the history of energy: centered around coal [Enerugīshī kenkyū: sekitan o tyūshin toshite], vol 32, pp 1–14

***SMART AGRI's Editors (2019a) Is production in "plant factories" the ideal of farming? Present situation and future tasks". https://smartagri-jp.com/smartagri/157. Accessed 20 Jan 2023

***SMART AGRI's Editors (2019b) The reason behind the attention riveted to the "partnership" or fusion between agriculture and welfare. https://smartagri-jp.com/agriculture/540. Accessed 20 Jan 2023

***Suezaki Ē (2018) An analysis of essential target recognition in social welfare: in the pursuit of historical and sociological inevitability recognition and the search for its quality of law [Kon-nichi no shakaifukushi ni okeru honshitsutekitaishōninshiki no bunseki: rekishitekishakaitekihitsuzensē no ninshiki to hōsokusē no hakken o motomete]. Bunnrikaku, Kyoto

**Teruoka S (ed) (2003) Japanese agriculture 150 years: 1850 to 2000 [Nihon no nōgyō 150 nen: 1850–2000 nen]. Yuhikaku, Tokyo

***Tōi M (1963) Our new agricultural administration and "economical utilitarianism": the keynote of the Agricultural Basic Law [Waga shin-nōsē to "kēzaigōrishugi": nōgyōkihonhō no kichō]. Econ Rev Kansai Univ 13(1–2):21–42

***Tsuno Y (1995) Small agriculture: a quest from mountainous farm villages [Chīsai Nōgyō: Sankanshinōson kara no Tankyū]. Rural Culture Association Japan, Tokyo

van Gerven M, Malava T, Saikku P, Mesiäislehto M (2024) Towards a new era in the governance of integrated activation: a systematic review of the literature on the governance of welfare benefits and employment-related services in Europe (2010–21). Soc Policy Adm 58(3):329–343

*Yoshida Y (2019) The situation and the task of farming with people with disabilities under the shortage of agricultural manpower. Agric Mark J Jpn 28(3):11–21

Chapter 2
The Significance of Agriculture-Related Job Opportunity Creation

Abstract Creating job opportunities as a solution to social problems is a practical method of social inclusion. This approach informed the action research project I initiated, which aimed at farmland recovery and maintenance through the involvement of homeless elderly laborers and entrepreneurs assisting youth employment. I conducted interviews with participants to examine whether participants view these job opportunities as more than simply a response to labor shortages. The homeless elderly laborers, who sought jobs that they can start remaining homeless, sometimes served as leaders to the young trainees. Farmwork, as part of employment assistance for jobless youth, not only boosts trainees' self-esteem but also helps them identify their aptitudes, making it an effective preliminary step in their job-hunting activities. However, some trainees expressed doubts about using farmwork as a means of adapting to the existing labor market. An additional inquiry examined the tension in introducing farmwork into employment assistance for these young trainees. The most common sentiment expressed in the interviews indicated that they were deeply absorbed in their work, which may be because the necessities and meanings of every detail in agriculture are governed by the laws of nature, which humans cannot control, unlike social or cultural norms.

Keywords Action research · Agricultural labor · Employment assistance · Nature · Sense of liberation

2.1 Introduction

If a social enterprise seeks both social responsibility and business sustainability, creating job opportunities through its professional activities may be a viable strategy (cf. Nakayama and Hashimoto 2006; Hashimoto 2007; Iwata 2008). Given the labor shortage in Japan's agriculture, which has resulted in an increasing area of farmland left uncultivated for decades, farmland recovery and maintenance projects

---

The original text written in Japanese, Tsunashima (2015), underwent major revision for the present chapter.

H. Tsunashima, *Farming and Social Inclusion in Japan*, Urban Agriculture,
https://doi.org/10.1007/978-3-032-01598-3_2

could provide employment opportunities for individuals who are often excluded from the conventional labor market. However, such projects will not be sustainable as a means of social inclusion unless the laborers involved perceive the job opportunities as meaningful, beyond merely addressing the labor shortage. Practitioners must seek feedback from these workers.

In this context, I initiated an action research project aimed at creating job opportunities in a suburb of Osaka City. I invited homeless elderly laborers and providers of employment support services for youth to participate in recovering an uncultivated plot of farmland. This raises several questions: what motivated the participants to join the project, did they feel that it met their needs, and if so, what implications does this outcome hold? The goal of this chapter is to address these questions.

The structure of this chapter is as follows: Sect. 2.2 outlines the issues discussed throughout the chapter, beginning with an exploration of how farmland recovery has become a significant social issue tied to social welfare. Section 2.3 explains the motivations behind the project, providing a historical context of the study site and a comparison of the project's development process with the general methodology of action research. It also discusses how data were collected through participant interviews. Section 2.4 presents the results of these interviews, while Sect. 2.5 analyzes the significance of creating job opportunities through farming activities and the remaining unresolved issues.

2.2 Background: Farmland System and Social Welfare/Inclusion

First, I will clarify the background in which the rise of uncultivated farmlands became a social issue, in relation to changes in the relationship between farmland and cultivators in post-WWII Japan. It was not until the 1980s that Japan experienced a significant increase in uncultivated farmland, a consequence of the decline in the number of farmers (MAFF 2010). Prior to this, Japan's agricultural theory was based on the assumption that only farmers should engage in agriculture and that only those with professional training could offer technical support to farmers. In other words, non-farmers were excluded from the agricultural sector (Nakagawa et al. 2006). Consequently, the issue of labor shortages within individual farming households was seen as a personal problem (Mishiro 2012).

In 1989, the Farmland Use Enhancement Law was revised to promote the efficient use of idle farmlands, with lawmakers aiming to allocate land to farms seeking to expand their scale (Ogata 2013). This led to strict restrictions on farmland lease and ownership, though these restrictions were gradually relaxed (Akiyoshi 2011; Takebe 2011). Meanwhile, the Basic Law on Food, Agriculture, and Rural Areas, established in 1999, emphasized the multi-functional nature of agriculture, leading more people to view farmland as a public good (Nakagawa et al. 2006; Akiyoshi

2011). In 2009, the Agricultural Land Law shifted away from the post-war notion that landowners should be cultivators, instead stating that farmland is a limited resource but should be accessible to ordinary citizens. This revision included provisions to prevent farmland from becoming disused. The Ministry of Agriculture, Forestry and Fisheries (MAFF) began recommending that landowners consider new types of users, such as non-farming residents, enterprises, non-profit organizations, and urban residents, if no farmer was available in their neighborhood to take over idle land (MAFF 2010).

This shift in farmland management led to a growing recognition of "quasi-farming," which, while not focused on agricultural productivity, provides social participation opportunities, particularly for the elderly, by utilizing uncultivated farmlands (Matsumiya 2013). Additionally, since the mid-1990s, the introduction of "horticultural therapy" from Western countries has contributed to the concept of "agricultural production-oriented welfare." In this model, citizens first engage with agriculture through horticulture for well-being and later contribute to farmland maintenance or agricultural production (Toyohara et al. 2007). As a result, recent studies have focused on the welfare aspects of farmwork (Tanaka et al. 2011). At the same time, the welfare policy for people with disabilities has shifted from relief to employment assistance. Numerous studies have highlighted the challenges employers face when hiring disabled workers (Ushino et al. 2007; Yasunaka et al. 2009; Ōsawa 2010). This has fueled a movement toward the "partnership between agriculture and social welfare" (Kondō 2013), under the belief that the labor-short agricultural sector can offer employment to those excluded from the traditional labor market.

As noted above, a mutually supportive relationship between farmland policy and social welfare has emerged. However, the core objectives of both—addressing labor shortages and utilizing farmland for other purposes—are widely different. While previous research has examined the technical challenges of integrating laborers without farming experience, it has not critically addressed the idea of employing individuals excluded from the labor market to solve the agricultural labor shortage. A lesson was gained from a pioneering initiative for social inclusion. The Community Employment Support Project, implemented in Osaka Prefecture since 2002, demonstrated that job seekers have diverse ethical, social, and financial motivations (Investigation and Research Section of Osaka Personal Support Scheme 2012). While it is difficult to universally determine how much consideration should be given to these desires when creating new jobs, it is essential to evaluate whether the job opportunities are meaningful for the workers. Otherwise, we risk exploiting workers to meet labor needs. In this context, I created opportunities to work on a farm without targeting a specific population, aiming to understand who needs these opportunities, why, and whether they find intrinsic value in farmwork.

Finally, once a parcel of farmland is recovered but not maintained, it will return to the wilderness, becoming difficult to cultivate again. Thus, the act of recovering farmland is inseparable from its subsequent maintenance. Therefore, "recovery" and "maintenance" are considered merged concepts in this context.

2.3 Research Process and Methodology

2.3.1 Study Site

The study site is located in Jōtoku Valley, which lies between two villages, Aodani and Karindo-obata, in Kashiwara City, Osaka Prefecture. Historically, Kashiwara City was known for its high-quality cotton production and marketing, but by the 1900s, this industry began to decline due to competition from low-cost imported cotton. During this period, grape production expanded and gained recognition as the region's new specialty. Proximity to the large consumer market in Osaka contributed to the success of the grape industry. However, by the 1970s, rapid economic growth and advancements in cold-storage transportation increased competition with other prefectures, such as Okayama, leading to a decline in grape cultivation. Additionally, the proximity to Osaka drove up labor costs, further accelerating the stagnation of grape production (Kodera 1986).

In the valley's lower section, villagers had developed terraced paddy fields on steep slopes long ago and cultivated wetland rice until the 1980s. These small land parcels, with winding ridges and narrow farm roads, hindered mechanization, making the introduction of rice-planting machines impractical. Consequently, these paddy fields were abandoned. Along with the decline in grape production, the rice farms faced challenges such as a lack of successors and an aging workforce, resulting in the degradation of irrigation channels, encroachment by wild plants, and crop damage from wildlife.

As a result, approximately 20 ha of farmland have remained uncultivated. This was seen as a problem by municipal authorities. In 2005, the municipality and the Kashiwara Social Welfare Council developed a plan to revitalize these farmlands (Okamoto 2005; Kashiwara City Social Welfare Council 2004). Although the plan was discontinued the following year due to difficulties in purchasing or leasing the land, it aimed to prevent further disuse and promote the land as a space where a diverse group of people could interact. Specifically, the plan proposed three goals: to provide opportunities for disabled individuals and senior citizens to engage in farm-related activities, to train young people interested in farming, and to offer work opportunities for individuals who had been excluded from the labor market for an extended period. Notably, the concept of a "partnership between agriculture and welfare" was already envisioned at this stage.

In 2007, the Osaka Prefecture government enacted the Ordinance Regarding the Preservation and Utilization of Farming Environments and the Encouragement of Urban Agriculture, which addressed the region's unique agricultural circumstances, including the smallest average farming scale per household in Japan. In this context, the Kashiwara City Farming Environment Preservation Committee designated Jōtoku Valley as a Zone to Take Precautions Against Farmland Disuse. The local agricultural administration office held a briefing for landowners to explain recent legal revisions aimed at curbing the growing trend of farmland abandonment. The new laws established a "right to use" farmland, enabling leases without transferring ownership. This cautious approach stemmed from the historical lesson learned by landlords during the post-war land reforms, which emphasized that once leased, land would not be returned. In a 2009 survey, landowners indicated that they would be willing to lease their uncultivated land, but only if tenants ensured the land would be returned without compensation for its release.

Following the 2008 financial crisis, the Japanese government allocated funds from the Special Emergency Fund for Job Creation to support the "Project for the Preservation and Utilization of Uncultivated Farmland" for the next 3 years. Under this initiative, the committee hired workers to mow the farmland in the valley. In October 2010, after plowing 3 ha of land, the committee hosted an event called "Action for the Preservation of Terraced Paddy Fields of Home," offering urban residents an opportunity to sow Chinese milk vetch. Shortly afterward, one of the landowners, referred to as X, approached me to discuss how to make productive use of his land. He expressed concern that if the land were mowed but left uncultivated, it would eventually revert to its original state (Fig. 2.1). In retrospect, X likely considered me a suitable candidate to manage his land, as I had been cultivating a small plot with various vegetables in a nearby location and had occasionally shared samples of my produce with him. Around that time, grants for scientific research were being offered, so I submitted an application outlining how I would create jobs using X's land. My proposal was accepted, and the funding enabled me to employ prospective participants in the project.

Fig. 2.1 Initial state of the project site

2.3.2 Methodology

When I began implementing this idea, my only assumption was that I would initiate a farmland recovery project to create job opportunities for individuals in need. I did not know precisely who these individuals would be. I anticipated that most job seekers would initially adopt a cautious, wait-and-see attitude. Therefore, I believed the project should be designed to allow for observation from the sidelines. Drawing on conventional action research methodologies (Susman and Evered 1978; Dickens and Watkins 1999), I adapted the traditional PDCA (Plan-Do-Check-Act) cycle to fit the context of this project, as follows:

1. The project begins with a small number of participants.
2. Progress is recorded and communicated to those observing from the sidelines.
3. New participants join the project.
4. Both new and existing participants assess the project's current state and identify new issues to address.
5. Participants collaborate to address these new issues.
6. Steps (2) to (5) are repeated, with increasing involvement from new participants and accumulation of experience.

2.3.3 Developments in the Present Project

I first involved Y, a staff member of an employment support organization focused on assisting young people. The wage for individual participants was set at JPY 3000 per half day, aligning with the municipal and prefectural job opportunity provision scheme in Osaka City (see Chap. 1), the "Special Scheme of Job Opportunity Provision for the Elderly." I then invited Z, a homeless elderly laborer I had known for several years. The operation began in the summer of 2011. Initially, Y's colleagues and their service users joined, followed by Z and his associate.

The service organization Y worked for, hereafter SO1, was established in 2007 as a Limited Liability Partnership with the mission of "offering a variety of job and education opportunities in supportive environments to people facing challenges in local communities."[1] SO1 organized both long- and short-term employment support programs: one with no time limit and another lasting 2 weeks. In both, users visited multiple workplaces to gain experience. As part of these programs, an instructor brought several users to the project site almost every week until the end of fiscal year 2012, March 2013. In total, 71 users participated in the farmwork. The frequency of their attendance varied, from a single visit to more than 40 sessions, depending on the instructors' assessments of how beneficial the farmwork was and the participants' eagerness.

[1] SO1 was dissolved in May 2014.

In the first 2 months, we worked only 1 day per week, focusing on farmland recovery tasks such as mowing wild plants, digging drainage channels, and rough plowing (Fig. 2.2). We then began cultivating vegetables on the recovered land. Once we achieved a stable harvest, we decided to sell the produce at a café in downtown Osaka (Fig. 2.3), a practice I had previously done. Selling our produce had two advantages: it provided a steady income to sustain the project, and it allowed SO1 instructors to offer their users a different kind of training than in typical SO1 programs. Consequently, we added an extra working day each weekend. Participants and I took turns serving customers.

As word spread about the project, new participants joined. Initially, several other service providers became involved as part of their own programs for young individuals at risk of employment difficulties. For example, SO2, which ran two different programs, provided farmwork opportunities to welfare recipients and senior-year students from part-time high schools. Additionally, Y and Z introduced other homeless elderly laborers and former homeless welfare recipients to the project. We also accepted instructors and users from an alternative school, hereafter AS1, which served adolescents who had disengaged from conventional schooling. Students from various universities also volunteered and provided significant assistance to the project (Table 2.1).

By October 2012, we had completed the recovery of the 3000 m^2 farmland (Fig. 2.4). We then cultivated the entire plot with vegetables, and occasionally

Fig. 2.2 Farmwork in the earliest stage of the project

Fig. 2.3 Selling of the farm produce

grains, maintaining at least 20 species at all times. While this approach was tentative, participants did not propose specific suggestions for further land use, despite my hopes for their full involvement. As I was the only one with cultivation experience, I ended up assuming the role of director, providing instructions to participants and taking responsibility for material supply and work schedule coordination.

By the end of March 2013, when I completed the interviews described below, the total amount of work was 1323 person-days. Of these, 496 person-days were contributed by young participants receiving employment support, and 244 by elderly laborers (Table 2.1). Labor costs for 304 person-days were covered by government subsidies for scientific research and farm produce sales.

2.3.4 Data Collection

For participants who frequently worked on the farm, I conducted semi-structured interviews. This interviewing technique allowed flexibility, enabling me to adjust the order and detail of pre-planned questions based on the flow of conversation. This approach encouraged open-ended responses, allowing interviewees to share information that might not have emerged from a more rigid question structure.

The interviewees included: Y, an SO1 instructor; 13 SO1 users who attended the farm more than ten times by the end of March 2013; two homeless elderly laborers, Z and another; five SO2 instructors; and three AS1 members. They were asked about how and why they joined the project, what they gained from it, and what they expected from it. After completing the interviews, I compiled the data into a single report, which was shared with the interviewees, sparking discussions via email. I also recorded the progress of these discussions.

Table 2.1 Intention and frequency of participation of participants from the opening to March 2013

Affiliation of participants (legal personality)	Purpose of organization or enterprise	Intention of participation	Frequency of participation (person-day)[a]
SO1 (Limited Liability Partnership)[b]	Employment support for the jobless youth	Job training	384 + 83
SO2 (Limited Liability Partnership)			
Step-mate Program[b]	Employment support for part-time and correspondence senior high school students	Job training	52 + 17
Osaka City's Job Attack Program[b]	Employment support for the jobless youth in welfare recipient families	Job training	19 + 4
SO3 (Social Welfare Corporation)	Employment support for the jobless youth	Job training	27 + 17
SO4 (General Incorporated Association)	Employment support for the jobless youth	Job training	14 + 1
Homeless elderly laborers, ex-homeless welfare recipients[b]	Personal	Employment	244
Former users of SO1[b]	Personal	Employment	83
AS1 (Incorporated Nonprofit Organization)[b]	Offering a place to belong for the youth who experienced social withdrawal or school nonattendance	Farmwork experience	37 + 25
Others	Personal	Farmwork experience	316
Total			1323

[a] The figures before and after "+" correspond to users and staff members including instructors, respectively

[b] Include the participants who gave interviews to the author

2.4 Interview Results

2.4.1 *SO1's Instructor and Trainees*

2.4.1.1 Overview of the Firm and Programs Concerned

SO1's programs cater to individuals excluded from both social welfare systems and the labor market. Many participants are referred by public employment support services, including Public Employment Security Offices, Youth Support Stations, Local Employment Support Centers, and Employment and Livelihood Support Centers for Persons with Disabilities. Service users often report that while counseling with these public providers is helpful, it leaves them unsure of how to take action or receive clear instructions, particularly for those with no prior work

Fig. 2.4 Completion of farmland recovery at the project site

experience. SO1 aims to clarify the strengths individuals can develop and the weaknesses they need to address in the workplace, which is particularly important for those diagnosed with developmental disabilities.

For example, a public organization officer asked Y to identify the cause of a client's repeated failures in job interviews. Y observed the client's performance as a trainee at a job training site and suspected of autism. This disability, if disclosed to employers, would not have impeded the client's success in job hunting. Instructors also receive various requests from prospective service users, such as: "I am recovering from social withdrawal and depression and want to re-engage with others," "I have made numerous attempts to find a job, but have been unsuccessful," "I've had jobs in the past, but they didn't last. I'm uncertain which job suits me," and "I have multiple disabilities, and I need assistance in identifying employment barriers."

Intake interviews address a broad range of challenges encountered during job searches and after employment. Meanwhile, SO1 works to develop a "sheltered" labor market by collaborating with companies that are willing to offer trainees on-the-job training. These trainees, who have been excluded from or need distance from the mainstream labor market, receive training in these supportive enterprises. Instructors observe both strengths and weaknesses in trainees that may not be apparent in other environments. Based on these observations, instructors guide service users in understanding their own characteristics. As the program nears completion, clients are referred to relevant support organizations, when applicable. For clients who wish to conceal their disabilities while job hunting, instructors encourage them to devise strategies to manage their challenges independently, as prospective employers may lack understanding of their difficulties. Some users begin job hunting or independent training midway through the program.

SO1 provided a stipend to users working as trainees in 2011, funded by the Special Emergency Fund for Job Creation. After the program's conclusion, some users continued training without compensation. These individuals may have had issues unrelated to job hunting, such as living environment problems, waiting for support from an Employment Transition Support Office, or still needing to improve attention concentration or social awareness.

2.4.1.2 Reason for Joining the Present Project

Initially, SO1 offered job training in partner companies or through staff contacts, which highlighted the need for self-managed and self-sustained training environments. This insight led SO1 to apply for the government's Special Emergency Fund for Job Creation in 2011. As part of this new initiative, SO1 ventured into agriculture before joining the current project, using previously uncultivated farmland in different neighborhoods to provide job training. By January 2013, almost all users had somehow experienced farmwork.

2.4.1.3 Achievements Instructors Realized

Instructors found that farmwork, in general, is well-suited for job training, particularly for two key reasons. First, it allows instructors to observe most trainees' cognitive, operational, and behavioral characteristics, helping identify tasks for further development. Second, farmwork helps trainees immerse themselves in tasks, which boosts self-efficacy and physical strength. A notable feature of farmwork is its ability to help trainees who struggle for attention concentration to overcome such challenges. Finally, unlike factory or retail environments, farmwork offers flexibility, allowing instructors to adjust the complexity and demands of tasks to meet the needs of individual trainees.

For instance, one trainee with significant cognitive challenges often became distracted and struggled to begin tasks, such as weeding a large area. However, once the task was divided into smaller sections, the trainee successfully completed the work. Based on this experience, the instructor advised the trainee that, should he wish to job-hunt without disclosing his autism, he would need to take similar steps to address his challenges independently. Another trainee with low self-esteem tended to avoid tasks in which he lacked skill. The instructor used farmwork to break tasks into manageable steps, allowing the trainee to see clear goals and complete the work smoothly. These small successes gradually expanded the trainee's capabilities. Similarly, a different trainee initially resisted tasks such as ridge-making or using a hoe. Upon recognizing that the trainee excelled at imitating others, the instructor demonstrated the proper technique for holding a hoe, leading the trainee to successfully complete the task and enjoy physical activity.

2.4.1.4 Achievements Clients Realized

All 13 interviewees had either been diagnosed with a developmental or intellectual disability or had experienced social withdrawal. Their work experience varied; some had worked full-time, while others had never held even part-time jobs. The reasons for their admission to SO1 were consistent with the instructor's prior explanation, particularly for Trainees A and B, who required preliminary treatment before following prearranged training programs focused on skill acquisition and providing work-experience opportunities (Table 2.2). Only four interviewees (Trainees C, G, J, and L) chose SO1 over other employment support organizations due to an interest in farmwork; the remaining interviewees joined without any specific interest. No clear relationship was found between interest in farming and the likelihood of securing agriculture-related jobs.

As noted by the instructor, several interviewees connected their understanding of personal traits with their successes when asked, "How do you think you can use your farmwork experience?" For example, Trainee F learned "to stay calm" and focus on executing tasks positively, even when reprimanded for mistakes, which had previously caused irritation. Trainee G struggled with a short attention span but developed strategies to maintain focus. Reflecting on his experience, he remarked, "I came to understand what it feels like to achieve something by completing tasks one by one." Other interviewees gained insights into the full range of labor; for instance, Trainee B "learned how to learn a job," and Trainee F "began to consider what it is to consider" in his own way. One critical comment regarding the use of farming tasks for job training was: "I learned how to hold a hoe and which muscles to flex when tilling the soil. But unless I enjoy this type of work, the experience would be useless" (H).

The most common response was that farmwork helped them "gain [or regain] physical strength" (Table 2.2). Notably, six out of the seven interviewees who mentioned this had not initially wanted farmwork included in their training menu. This raises the question of why they continued to work on the farm, even though they had either lost physical strength from prolonged unemployment or were not strong enough for traditional employment. This may indicate, in line with the instructor's perspective, that the trainees were unconsciously engaged in certain aspects of farmwork.

To explore this further, I asked each trainee to categorize farm tasks they found "laborious only," "laborious but continuable," "laborious but worthwhile," and "enjoyable," based on their actual experiences. Table 2.3 compares these categories. Tasks that reminded trainees of their weaknesses, were extremely physically demanding, and yielded less immediate results were categorized as "laborious only" by Trainees A, H, and M; C, G, and J; and F and Φ, respectively.

Importantly, all interviewees found some farm tasks "worthwhile," and most also found them "enjoyable." Nearly half admitted that some tasks were "laborious but continuable," indicating that immersion in farmwork was possible. Trainees expressed the experience of absorption in their work as follows: "I couldn't escape

Table 2.2 Significance of training conducted in the farm from the standpoint of SO1 users[a]

User (age)	Motive for the use of SO1	Achievements of training conducted at the farm[b]	Future prospects
A (20s)	A vocational aptitude test indicated that I would be most suitable for the job I had recently quit because I made many mistakes. So, I was at a loss as to which job I was suitable for when I saw a leaflet on SO1	I gained in physical strength. Now I am working at construction sites as a part-timer, and I think nothing of such heavy labor. I used to have pent-up feelings because I was still looking for a job and keeping irregular hours, but only when I am engaged in physical labor, including farmwork, do I find such feelings of mine cleared up	It would be tough for me to make a living only by farming, but it could be a realistic option if I had a side job to work at night as a part-timer near the farm
B (20s)	When I was using another employment assistance organization and receiving a workplace adaptation training program, my strength ran out before the finish. Then, a staff member of told me to experience farmwork elsewhere to improve my physical strength	• I gained in physical strength. Now I am strong enough to work 8 h a day. Not only physical labor but also clerical work requires a certain level of basic strength • While doing farmwork, unlike clerical work, it was impossible to take notes of how to perform each task because I often got my fingers soiled, and I had no choice but learned how to do it just by doing it. In so doing, I learned how to learn a job	If it were possible, I want to make a living by farming, but I know it would be hardly realistic. I have made no further decision
C (20s)	After the employment period of my then part-time job ended, I began to search for a job in the field of agriculture. I then chanced on recruit information from SO1	• I gained in physical strength and grew accustomed to rising early • I have come to consider which materials and tools I have to prepare for each task	I want to be engaged in something related to agriculture. I will try to keep up the present physical strength and early habits. I will apply for jobs for non-disabled people, even though I may end up getting a mere part time job
D (20s)	Since I was diagnosed with a developmental disability, I applied for a job from SO1 for the disabled	• I gained in physical strength by doing manual labor • I am gradually realizing that, if someone tells me how to execute it and I get the knack of it, I can perform a variety of tasks	I will try to obtain a regular job in other fields than agriculture

(continued)

Table 2.2 (continued)

User (age)	Motive for the use of SO1	Achievements of training conducted at the farm[b]	Future prospects
E (20s)	An instructor of a vocational-training school recommended that I see a doctor, and then I was diagnosed with a developmental disability. At that time, the doctor mentioned SO1	I realized that I was physically stronger than other participants when I compared myself with them after receiving the same training. That gave me confidence in this regard	I hope to work for a manufacturer
F (20s)	When I was diagnosed with a developmental disability, the psychiatrist introduced SO1 to me, saying that the situation would improve if I dealt with people similar to me	• I gained in physical strength. As I take the job training in a quiet environment which allowed me to stay calm, I developed a habit of considering how to execute the present task even when reproved by my superior or instructor. Otherwise, on an occasion like this, I would unnecessarily get irritated instead. Now I have learned to suppress my feelings and enter into those of the person before me • I also developed a habit of pondering by myself, for example, how I should hold a hoe to save my strength. The point is the interpretation of what I am told to do instead of the mere performance of it • In the end, I began to consider what it is to consider	Now I am temporarily working in the distribution industry. I hope to be engaged in a job that similarly accompanies manual labor
G (20s)	I chose SO1 from several employment assistance enterprises because I was interested in farmwork	I came to understand what it feels like to achieve something by completing several tasks one by one. Formerly, I could focus on the task I had to perform only when I was listening to music using headphones. However, presently, I can do that if only I tell myself, "Bother about nothing else". I hammered out this method by myself	I hope to work in the fashion industry
H (20s)	One of my parents was acquainted with one of the SO1 staff	I learned how to hold a hoe and which part of my muscle to flex when tilling the soil. However, unless I liked to do such work, this experience would be useless for obtaining other kinds of job	I will try to obtain a job after I have received additional training at an Employment Transfer Support office

(continued)

Table 2.2 (continued)

User (age)	Motive for the use of SO1	Achievements of training conducted at the farm[b]	Future prospects
Φ (30s)	After consulting several employment assistance organizations, I was finally referred to SO1	I had lost a lot of strength at the time when I began to take training in SO1. However, I have regained it	I hope to work as an intern for a farm or a business in agriculture-related industries
J (30s)	When I joined a session offering farmwork experience, I met a person concerned with SO1, who invited me to consult SO1	• I learned agriculture from the very basics, for example, how to sow seeds • The farmwork experience is unlikely to be useful for overcoming the problems I have been facing in finding employment	I want to be a childcare worker and do farmwork together with children
K (30s)	I was referred to SO1 at a Public Employment Security Office	I gained in physical strength, of which I will somehow make the most, if not immediately	I hope to be engaged in farmwork while using Continuous Employment Support
L (20s)	When I chanced upon a job-offer leaflet of SO1 and made contact, I said that I have never worked for a ranch. A staff then member invited me, saying, "You should try farmwork, because you have some experience"	• Since the present farm requires a lot of handwork, I enjoyed working together with many fellows • A variety of experience, from seeding, daily maintenance to harvesting, will be useful hereafter	Now, I am searching for an agriculture-related job in Public Employment Security Offices
M (20s)	After consulting several employment assistance organizations, I was finally referred to SO1	Farmwork experience will not be very useful to me	I do not want to be engaged in farming because I am not strong enough

[a] Personal information is partly obscured
[b] The phrases underlined are cited in the text

from becoming unnecessarily particular about the outcome of my task" (Trainee A), "I can sustain single-minded concentration" (Trainee H), and "time passes without my noticing while working" (Trainee L). Other phrases such as "I enjoyed it without knowing why" (Trainee E) and "though I don't know why" (Trainee M) suggested that the reasons for their enjoyment and absorption were difficult to articulate.

Additionally, although some interviewees considered certain tasks to be "laborious only," they still found farmwork to be worthwhile and enjoyable enough to offset the effort involved. For example, Trainees B, G, and Φ stated, "Although some tasks are still tiring, I do not feel exhausted in the same way as when serving customers at a storefront," and described the excitement of "seeing the full posture of vegetables" and "observing various creatures."

Trainee A discontinued his involvement with SO1 and obtained a job at a regular company. After several months, he was hired as a staff member of the farm and subsequently received full-scale training to become an independent farmer, supported by the MAFF's subsidy for new farmers at the start of fiscal 2014.

Table 2.3 Classification of farm tasks from the subjective viewpoint of SO1 users

User	Laborious only	Laborious but continuable[a]	Laborious but worthwhile	Enjoyable[a]	Related remarks[b]
A	• Weeding because I dislike spiders	None	• Leveling the ground because I had a feeling of achievement when pathways of a cultivator or waterways were completed • Harvesting leguminous crops because it allows participants to make decisions on which pods to pick or leave, and it yields a lot of harvest at once. I enjoy seeing many packaging bags packed with pods lined up in rows when preparing for selling	• Sloping of furrow to facilitate drainage. I could not escape from becoming unnecessarily particular about the outcome of my task, which coincide with the fact that I tested well at space perception in a vocational aptitude inspection	
B	• Serving customers	• All farm tasks because, although it is still tiring for me to conduct them in a sense, I do not feel exhausted at all in the same way I do when serving customers at the storefront • For example, the above holds true for transplanting garlic partly because, to be specific, it is indeed tough for me to keep squatting down for a long time in the hot weather	• Digging drainage ditches because I would feel happy to see water drain away • Ridge making	• Digging drainage ditches and building ridges because it is easy to see what to do • Soil preparation because it is exciting to start developing the land pieces that have been left uncultivated for many years	• Farming environment allows me to work naturally in an unrestrained manner
C	• Working in the situation where agricultural machinery is used because its engine noise distracts me from the task	• Rough plowing because I am better at rough tasks than precision ones, and, still, I am nonathletic but I will improve in performance if I am conscious of this shortcoming	• Ridge making because I get a sense of achievement when we have finished to develop the entire area of the present farm into cultivable land	None	

D	None	• Relaying bamboos on a slope after thinning out the forest because I performed this task quite successfully except for a hot season • Manual labor work because I was growing accustomed to conduct as I gained physical strength	• Transportation of bamboos because I enjoyed it as good exercise • Ridge making, because I was eager to use hoes since I had attained proficiency	None	• Although I prefer the repetition of the same tasks, even the experience of frequent changes in task was valuable
E	None	Not specified	• Ridge making with a cultivator because I actually did it	• Use of a cultivator because <u>I enjoyed it without knowing why</u> • Logging timbers for mushroom cultivation because I had learned how exciting it is to shape my idea into a thing	
F	• Fallen leaf collection for manure making in rainy whether in winter season because I would get my hand cold and hardly have a sense of accomplishment	Not specified	• Producing goods in cooperation with other people • Batching and packaging after harvesting of vegetables • Cooking and tasting harvested vegetables • Soil preparation because, once we do that for the present land parcel, someone will be able to readily grow crops	• Sowing seeds • A part of transplanting, i.e. make a hole in the ground and put seedlings in the hole • Tilling with a hoe, because I got the knack of it	• I realized that I could work in a quiet atmosphere • I feel comfortable when I touch the soil

(continued)

Table 2.3 (continued)

User	Laborious only	Laborious but continuable[a]	Laborious but worthwhile	Enjoyable[a]	Related remarks[b]
G	• Working in hot weather • Tasks such as weeding, which we have to crouch to perform	• Digging the ground for some purpose	• Harvesting because all the product will be tasted by someone	• Ridge making • Moving around in the farm because I can <u>see the entire posture of vegetables</u>	
H	• Ridge making, because it is troublesome to adjust the height • Watering after sowing, because we have to carry a heavy watering pot • Tasks to be finished in haste, because I am apt to make more mistakes on such occasions	• Uprooting of bamboo grass	• Weeding and uprooting of bamboo grass because I feel refreshed when the ground is cleared	• Weeding because it is easy to perform and <u>I can sustain single-minded concentration</u>	
Φ	• Removing weed roots from the inside of ridges because we could find them everywhere and thus hardly offered a sense of accomplishment • Street selling because we have to make a steady and sober appeal to passers-by to attract them with different emphasis depending on the item	Not specified	• Ridge making, because we have to use hoes, which makes me feel as if I have become a farmer	• Almost all tasks, unless they cause me lower-back pain, because I could <u>observe various creatures</u> living in the field especially in the summer season	• Farmwork allows us to work in nature, where we can make contact with creatures, breath fresh air, and admire the scenery

J	• Working in dirt because it is exhausting to take even a step forward when we get stuck in the mud with our boots on	Not specified	• Almost all tasks because we can taste what we have grown, which actually tastes well	• Sowing • Harvesting	
K	• Tasks such as digging drainage channels, which accompanies heavy manual labor	Not specified	• Almost all tasks because we can observe the crops grow and then reap them	• Harvesting	
L	None	• Digging drainage channels because it is exhausting to move the earth mainly composed of heavy clay, but <u>it makes me realize that time passes without my noticing while working</u>	• Harvesting and selling, because we can have the unsold, which tastes well	• Harvesting • Shaving the earth with a hoe [author's note: to ease drainage between ridges] • Selling because it is more enjoyable to wait on and deal with customers at the storefront than receive other forms of job training such as building cleaning	
M	• Felling bamboos because it is difficult to cut them sharply	• Mowing along drainage channels because I finished this task even though my boots were broken and my socks came off	• Almost all tasks because I felt that I worked until I got tired, which makes me sleep well, contrary to other forms of job training	• Weeding, sowing, and transplanting because they are enjoyable, <u>though I do not know why</u>	

[a] The phrases underlined are cited in the text
[b] Answers to other questions than those regarding to farm tasks "laborious only", "laborious but continuable", "worthwhile" and "enjoyable"

2.4.2 *Staff Members of SO2, Who Supported Part-Time or Correspondence High School Students*

2.4.2.1 Overview of the Firm and Programs Concerned

Among the primary activities of SO2 is providing employment assistance to youth. The organization primarily targets part-time or correspondence senior high school students, many of whom were former dropouts from full-time schooling and often face difficulties finding employment. In fiscal year 2011, SO2 was entrusted with the "Step-mate Program," a tutor training initiative aimed at supporting the employment of these disadvantaged students. This program was part of the "Osaka Prefecture's Emergency Job Creation Scheme (Priority Areas)."

The program's staff, including interns, performed several key tasks. First, they identified students struggling to envision their futures after graduation and those who lacked the drive to actively pursue career opportunities. These students were often stagnating under traditional support methods, such as career guidance classes and job matching. Second, if students were unable to determine a career path by graduation, staff members helped them find alternative paths outside of school.

Some students were still undecided about their post-graduation plans, even just months before graduation. The staff selected seven students who appeared to need special assistance and who agreed to participate in the "One-step Challenge Project." From February to March, when senior high school students typically have no school, these students received basic vocational training. These students often faced challenges such as school refusal or family issues, including poverty and unstable domestic relationships—problems that were difficult for both the students and their teachers to manage. Although they could not continue their education or remain idle at home, they had trouble in beginning on job searching. Additionally, their family backgrounds often excluded them from opportunities to visit their parents' rural hometowns or participate in school events, which might have provided valuable experiences like farmwork.

During the project, the students visited the farm four times with three staff members. On the remaining days, they participated in workplace experiences and attended lectures on relevant topics. Participants were given a daily allowance of 2500 yen for their involvement.

2.4.2.2 Reason for Joining the Present Project

One of the targeted students, who expressed an interest in agriculture, had already been visiting the farm weekly since November 2011, guided by a staff member during her free time in the mornings and early afternoons. Noting her positive response, the staff decided to take all the targeted students to the farm during the program term.

2.4.2.3 Achievements Staff Members Realized

As of January 2013, four of the seven targeted students had not yet found courses to pursue after graduation. However, once brought to the farm, they surprised the staff. The abovementioned student transformed into a different person, displaying active engagement and frequent laughter. Moreover, she managed several challenging farm tasks with patience and care. She also gained enough physical strength to reduce her commute time from 40 min to just 20 min—walking up an incline from the nearest railway station to the farm—an improvement attributed to her sense of obligation. "It was tough to commute, but I felt obliged to do so because I had insisted on going there," she explained.

On one occasion, I observed her happily performing a farm task with a smile, accompanied by an interviewee, who supervised her that day. The task, treading wheat plants (Fig. 2.5), is traditionally believed to promote tillering and prevent detachment from the soil due to hoarfrost. The interviewee later remarked that this rough duty brought an unexpected sense of refreshment.

Other students also showed remarkable personal growth during the program. One, previously interested in only a narrow range of jobs outside of farming, began to take on detailed work while actively guiding other students. He later expressed a desire to gain more experience in farmwork and ultimately completed an internship at a dairy farm. Another student, once quiet and reserved, unexpectedly engaged in conversation with an elderly participant while working.

The staff members had hoped to establish a relationship with the target students based on equal footing, though the students may not have shared this sentiment. Both the students and staff, who had little prior farm experience, found themselves

Fig. 2.5 Participants performing the farm task called "wheat plant treading"

starting at the same point. When a staff member made a mistake, the students would feel affinity with them. The relationship evolved further when, in situations where the students had already learned a task, the students occasionally provided guidance to the staff, which is uncommon in other workplaces. One staff member noted, "There is nothing contrived in the farming environment."

By incorporating farmwork into their workplace experience, alongside traditional urban industries like manufacturing and services, the students realized that "business manners" and other norms are neither universal nor absolute. This allowed them to expand their horizons and express themselves without stress or fear. Only after these changes could the staff identify each student's interests and aptitudes, enabling effective career counseling. In short, the restoration of self-esteem empowered the students to take the next step in their employment journey, even if not immediately seeking a job.

2.4.3 *Staff Members of SO2, Who Supported Jobless and Welfare-Recipient Youths*

2.4.3.1 Overview of the Programs Concerned

In fiscal year 2012, SO2 was partially responsible for the "Osaka City's Job Attack (Youth Employment Assistance) Program," a component of "Osaka Prefecture's Creating the Emergency Job Creation Scheme (Priority Areas)." The organization was tasked with assisting 10 welfare recipient youths. The program consisted of two branches: "Manufacturing" and "Environmental Engineering." Over 4 months, the trainees gained experience in various real workplaces, followed by onboard training at the companies where they sought employment.

The environmental engineering branch targeted five trainees who, despite being diligent, had little workplace experience and had failed to secure employment after participating in the foregoing "Try-and-Learn-Workplace Program." In this following program, staff provided on-the-job training at the farm once a week from May to June 2012. On other days, the trainees received similar training to the target students discussed earlier.

2.4.3.2 Reason for Joining the Present Project

The "environmental engineering" branch allowed staff members to freely design training content, which included farmwork to encourage physical activity in nature and enhance trainees' physical strength. It was observed that trainees often displayed a passive attitude, possibly because they expected someone else to assist them in times of difficulty. Despite being adept at following instructions, they were generally unwilling to take initiative, likely due to overprotective parenting that limited their autonomy. Some trainees struggled to express their needs, such as saying, "I need to rest," even when exhausted.

2.4.3.3 Achievements Staff Members Realized

None of the trainees secured employment with the companies to which they had been assigned. However, immediately after completing the program, one trainee obtained a full-time job, and another secured a part-time position. Additionally, two others made significant progress: one received a disability certificate, while the other began living independently—both considered important steps toward employment by staff members.

These achievements can be attributed to the farm's open-air, spacious environment. Trainees felt comfortable entering the farm, as other participants were absorbed in their own tasks and not focused on them. This contrasted with other training sites, where trainees often felt observed due to the confined, indoor setting. On the farm, trainees completed their tasks calmly, even with insects and worms around them, which they found distasteful.

At the beginning of the program, a 32-year-old trainee did not initially integrate well with younger trainees in their early 20s. However, as work progressed, the younger trainees often gathered around him after noticing his hard work and realizing their shared responsibilities.

Trainees typically avoided strenuous tasks in their daily lives. Similarly, at other training sites, they often stood motionless when uncertain about how to complete a task. In contrast, at the farm, trainees learned to independently assess what needed to be done, even when faced with physically demanding work. This fostered communication and collaboration among them.

Occasionally, staff members and trainees struggled to understand the details of instructions from the director. For example, when asked to sort onions by size (Fig. 2.6), staff encouraged trainees to begin the task despite possible misunderstandings, as they could adjust their approach later. While staff members may occasionally need to interpret instructions for trainees, it was more important that both staff and trainees gained valuable experience through trial and error.

In other training sites, where trainees often relied on staff for guidance, staff members were positioned above them. At the farm, however, staff and trainees had to collaborate and exchange knowledge, as staff could not provide instructions autonomously. This led to a dissolution of the hierarchical relationship between the two groups.

Moreover, farm tasks required trainees to navigate uneven terrain and exert more physical effort than they were accustomed to, which ultimately increased their physical strength. These changes allowed each trainee to reveal a side of themselves that staff members had not seen in indoor training settings. Reflecting on the trainees' progress, staff members assessed which training providers would be best suited to help them take the next steps in their job search. In conclusion, staff members concluded that farmwork offers the most significant advantages at the earliest stages of employment support programs.

Fig. 2.6 Participants sorting onions by size

2.4.4 The Elderly Homeless

2.4.4.1 Reason for Joining the Present Project

Many elderly homeless individuals insisted that, as long as they remained healthy, they would continue working rather than rely on welfare. Despite facing significant challenges in securing typical employment—especially construction jobs—due to both their homelessness and advanced age, they persevered. This was the case for Z and other participants. Meanwhile, the municipal government of Osaka City and Prefecture implemented the Special Scheme of Job Opportunity Provision for the Elderly. Through this initiative, Z and his companions were employed by the day for 6 days a month as street sweepers or mowers. This employment allowed them to barely meet the minimum standards for homeless life, although their earnings were less than half of the welfare benefits they could have received. As a result, they sometimes needed additional sources of income for extra expenses.

Initially, Z and his colleagues joined the current farm project for the modest, but reliable, wages. However, they discovered that farmwork offered greater appeal than the typical urban informal labor they had previously done, as evidenced by statements such as:

> We can breathe fresh air.
>
> Our legs get tougher.
>
> The farm tasks, ranging from seeding to harvesting, are more varied and exciting than street cleaning.

> The work pays reasonably well, even in the hottest season, especially when performing heavy tasks like mowing grass and cutting bamboo. We can also harvest crops we planted ourselves. I would like to continue working on the farm, in addition to my job under the scheme, as long as I am healthy, even when I reach 70.

When they had spare time but insufficient funds for train fare, they searched through garbage disposal spots for empty aluminum cans to exchange for money. However, one worker noted that when bending down for tasks like weeding and harvesting, they had to adjust their posture to reduce strain on their lower backs. Their primary focus was maintaining consistent attendance in the municipal job program. He emphasized that working on the farm was secondary to the work provided under the aforementioned scheme.

According to Z, he often observed his co-workers under the scheme enthusiastically engaging in tasks like weeding, showing nostalgia for their rural backgrounds. This suggested that, if the farmwork were financially more rewarding, there would likely be many potential participants. However, the current wages were insufficient unless one had a strong passion for farming. Additionally, the farm's work schedule conflicted with that of recyclable waste collection in suburban Osaka, during which homeless collectors could earn up to JPY 6000 a day. Some workers, who were entirely devoted to recyclable waste collection, did not appear suited for farmwork. For instance, another elderly homeless individual told me that, although Z had invited him to join the farm, he could not participate due to his daily commitments to other jobs.

2.4.4.2 Achievements Other Participants Realized

The previous section established that the farm was an attractive source of supplementary income for Z. Other participants also found it valuable to work alongside elderly homeless laborers. Z and one of his colleagues had limited experience in physical labor due to their prior office jobs until the 1991 economic bubble burst, making them exceptions among the elderly homeless. However, they both came from rural backgrounds. Having observed their parents' rural work as children, they quickly grasped how to use agricultural tools and accurately identified their purposes, even when encountering them for the first time. For example, they were able to distinguish between various types of hoes. Their extensive life experience and confidence also allowed them to teach younger workers how to use agricultural tools and collaborate effectively on tasks that required teamwork.

Many young people had never met homeless individuals and often held misconceptions, such as believing they *were* dangerous. However, through conversation and collaboration, they built positive relationships. One trainee (SO1 user, Trainee A) shared, "At first, I felt uneasy working with homeless people, and a friend asked me if it would be dangerous. But I quickly realized they are kind people." Another (Trainee F) noted, "I was initially reluctant, but when I met them, they were friendly and showed me how to carry materials for the fence and hammer posts into the ground (Fig. 2.7). This was helpful." He attributed his achievements, as detailed in

Fig. 2.7 Collaboration between an elderly homeless laborer and a young trainee

Table 2.2, to their active engagement in fostering relationships. Other SO1 users also expressed gratitude and respect for the elderly homeless laborers, describing them as more supportive than their employment instructors. One remarked, "It was easier to work with them than with our employment support instructors. When the latter scolded me and I felt down, the former cheered me up." Another stated, "When we took breaks from farmwork, the elderly homeless shared their life stories, which I found tough and attractive."

Despite this, Z and his colleagues were skeptical about being allowed to instruct young trainees, as the public often attributes homelessness to personal failures. They approached the young trainees modestly, which further highlighted their potential to guide individuals with little work experience. New job opportunities could arise from this dynamic. For instance, they inquired about how to manage trainees who appeared excessively sensitive. Initially, they expressed frustration with some trainees' seeming lack of motivation. However, after receiving an explanation from an instructor about developmental disabilities, they understood that these trainees were, in fact, highly motivated but overly sensitive and easily distracted. To fully realize the employment potential for elderly homeless individuals, further investigation is needed to determine which aspects of employment assistance can be effectively handled by them.

2.4.5 AS1

2.4.5.1 Reason for Joining the Present Project

The policy of AS1 is to provide a foundation for youth who have experienced social withdrawal or school nonattendance. While some participants ultimately earned enough money to sustain themselves—such as through caregiving—organization members firmly believe that unsolicited employment assistance often only serves to exhaust the individuals it aims to help. The involvement of certain staff members and users in the current farm project stemmed from their shared belief that alternatives to wage labor should be explored. Although many organizations had already incorporated farmwork into their programs to help withdrawn youth reintegrate into society, staff and users viewed this approach as overly focused on group labor to foster sociability for social rehabilitation. This approach conflicted with their core objection. They realized, however, that a different approach could be implemented on the current farm, as mentioned above. Specifically, given the high cost of basic living expenses in the current economic climate, they wanted to produce some of their own necessities—such as food—partially or entirely, rather than relying on wages to purchase everything.

2.4.5.2 A Question Raised After the Interviews

During the farmwork, most SO1 trainees appeared passive. It seemed questionable whether they were truly interested in farmwork, or if they were merely brought to the farm by the instructor. They began to doubt, and even resent, the employment assistance provided by SO1. As a result, one student, after their first visit, decided never to return; another, who had planned to attend with schoolmates, felt reluctant to join and ultimately was absent that day.

Based on the interviews summarized above, a staff member raised the following skepticism:

> Although unskilled youth are often excluded from the labor market, the issue should not be the youth's quality as a labor force, but rather the quality of the employment available in the regular labor market. The present farm not only offers agriculture as an alternative career path but also provides training for those hoping to secure jobs in the conventional labor market. It is true that SO1 trainees gained physical strength, which would support future employment. In this sense, the farm has proven effective in enhancing employability. However, is the purpose of farmwork merely to help trainees gain the physical strength needed to endure the deteriorating conditions of employment?

This skepticism suggests that uncritically promoting farmwork as an employment assistance tool at the present farm may alienate potential participants, leading to their eventual disengagement. An instructor from SO1, Y, responded as follows:

> As for the job training provided by SO1, the main goal of incorporating farmwork is to identify the characteristics of each participant that hinder their social adaptation. They should be encouraged to understand these factors. It may be true that this form of employment assistance eliminates 'social maladaptation,' which can otherwise act as a driving

> force for social reform. I also agree that we should 'question the quality of employment,' or adapt society to accommodate maladapted individuals. Nevertheless, we must confront the reality that our users are trying to adapt to the current societal structure. Public employment support agencies that refer clients to us often highly appreciate our ability to assess their strengths and weaknesses through on-site training. Ultimately, I hope that participants learn to make a living. However, I do not expect any of them to pursue a career in agriculture at this time.[2]

The views of these two participants align in that both emphasize the need to "question the quality of employment" in the long term. However, on a short-term basis, the former advocates for farming as an alternative path for social adaptation, while the latter sees it as a means of social adaptation aimed at securing employment. In other words, the key difference lies in whether farmwork is considered a detour toward the conventional goal of employment assistance or something distinct.

2.5 Significance of the Present Project

2.5.1 Different Views on the Significance of Farmwork and Diversity of Will to Work

The interviews reveal that farmwork opportunities held different meanings for participants. The AS1 members sought to produce their own food as long as others were excluded from the current labor market. Elderly homeless laborers aimed to generate supplementary income while continuing to work despite their homelessness. The two SO2 programs focused on youth with limited work experience, aiming to build trust between clients and their surrounding communities, which helped identify their interests and aptitudes. In contrast, SO1 targeted youth with past or potential exclusion from the labor market. Instructors focused on identifying relevant factors such as developmental disabilities, aiming to help trainees cope independently with probable workplace challenges. These efforts ultimately led to increased self-esteem and physical strength in trainees. However, this success reflects an approach that adapts trainees to the existing labor market. As noted in the previous section, AS1 members question the use of farmwork for this purpose.

The varying perspectives on the significance of farmwork parallel the diversity in participants' willingness to work. Meanwhile, the central government has implemented the "Self-Reliance Support Policy for Jobless Youth," emphasizing the "enhancement of motivation to work" (Satō 2008). However, there is no social consensus on the definition of "work" in this context (Kudo 2013), making it impossible to clearly define the phrase "enhancement of motivation to work." The government has defined this term from its own perspective, using the expression "lacking motivation to work" to refer to "not taking any forms of specific action toward job hunting" (Itō and Kimichi 2004). Thus, the government regards any effort made prior to job hunting as indicative of "motivation."

[2] This remark was offered before an ex-SO1 trainee decided to become an independent farmer.

Since SO1 and SO2 trainees were indeed at a standstill in their job hunting, the term “lack of motivation” aligns with their then current status. However, instructors did not view their trainees as lacking motivation. Instead, they sought to diversify training by incorporating farmwork, thereby tapping into the potential motivation of trainees, which could eventually lead to specific actions toward job hunting. Generally, many organizations provide job training to evoke and sustain motivation to work while facilitating skill acquisition (Kanai and Shikata 2013). This holds true for SO1 and SO2, whose programs were not necessarily aimed at acquiring skills for a specific field. Most participants in the current study, however, achieved improvements in physical strength, during which they became deeply engaged in farmwork. The implications of this finding will be explored in the next subsection.

Within the framework of the “Self-Reliance Support Policy” for jobless youth and the homeless, government officials assess whether a person has the “motivation to work” based on whether they have taken “specific actions toward job hunting.” This definition of “the will to work” excludes, for example, the desire to produce food independently without excessive efforts to obtain employment, as expressed by AS1 members.

What about agricultural administration? Is it possible for officials to recognize a form of “the will to start farming”? Until recently, authorities imposed a “minimum farming-scale requirement,” outlined in Article 3 of the Agricultural Land Law, to prevent “small-scale farming that lacks productivity, which hinders agricultural development and efficient use of farmland” (Agricultural Administration Office of Environment, Agriculture and Fisheries, Osaka Prefecture 2007). This policy indicates a significant gap between what authorities have considered “the will to start farming” and the desire of AS1 members to produce food. However, recent revisions since 2007, initiated by MAFF, have focused on encouraging retired salary workers to take over farming or start farming as a second career (MAFF 2007). Osaka Prefecture also launched the “Scheme for Enhancement of Semi-Farmers to Enter Agriculture,” which allows urban residents to lease small plots of farmland (about 300 m^2) and sell their produce. While MAFF continues to prioritize productivity, it has placed greater emphasis on involving non-farmers in farmland conservation, thus opening the door to recognizing the desire to farm for personal consumption.

Turning to the will to work among homeless laborers, the Ministry of Health, Labor and Welfare (MHLW) outlined the objectives of the “Job Assistance Scheme for the Homeless and the Like in Fiscal 2013” as follows:

> According to the results of the ‘Nationwide Fact-finding Survey on Homelessness,’ about 30% of homeless individuals who sleep outside indicated they would like ‘to live in an apartment and find a job to be self-reliant.’ This suggests that many people are homeless despite their desire to work. The government should implement support measures to facilitate self-sufficiency through employment and secure job opportunities for these individuals.

However, it is important to note that the questionnaire used in the survey (Investigation Committee for National Survey on Homelessness 2012) was based on multiple-choice questions. The answer cited above was one of the options for the question, "How do you hope to make your living from now on?" Other choices included:

2. To become a live-in employee and be self-supported
3. To become a welfare recipient due to inability to obtain employment
4. To live in an apartment receiving welfare benefits and find light work
5. To be hospitalized
6. To live with family members
7. To remain homeless, i.e., to stay as you are now
8. No idea
9. Other options

The "will to work" expressed by Z and others in the interview, which involved working while remaining homeless, does not fit into this framework. The researchers who designed the questionnaire seem to have overlooked this form of motivation, assuming that homelessness inherently reflects a lack of self-reliance, and remaining homeless was viewed as synonymous with lacking motivation to work. Consequently, analysts likely classified this as "to remain homeless." This assumption reflects the higher value placed on ending homelessness in the "self-reliance support policy" than on maintaining one's livelihood.

As discussed earlier, not engaging in "specific actions toward job hunting" does not equate to a lack of motivation to work. By prioritizing the "enhancement of motivation to work," the "self-reliance support policies" have overlooked several forms of motivation among jobless youth and the homeless who were already motivated to work in a broader sense. The interview results clearly show that participants' willingness to work takes various forms, and farmwork opportunities can support these diverse motivations. It should be noted that, in general, when addressing "people with lowered motivation to work," it is essential to examine the underlying values and ideologies (Japanese Association of Certified Social Workers 2010). The next question to consider is whether society will accept the desire to farm on a small scale for personal food production even unlikely leading to self-sufficiency, and whether it will acknowledge the will to work as an agricultural laborer while remaining homeless.

2.5.2 Essence of Farmwork Opportunities in Employment Assistance for the Youth

The phenomenon of immersion in farmwork proves significant within the context of social inclusion. Tsuno (1995) argues that when a person becomes absorbed in repetitive tasks as part of farmwork, they quickly reach the limits of their physical endurance and lose focus. However, at this point, their vision expands, and they begin to perceive their surroundings, including the crops around them, allowing them to gain a broader perspective on the outcomes of their labor. In other words, the experience of intense physical labor in a specific environment enables individuals to recognize the value of their efforts, fostering confidence in their ability to continue working. As a result, some individuals may begin to take specific actions toward job hunting. This is the change that the instructors of SO1 and SO2 aimed to inspire, and some users have indeed experienced it.

However, there is no inherent necessity to link the experience of immersion to such actions. As discussed previously, an ex-SO1 user abandoned the pursuit of an ordinary job and chose to become a farmer, despite coming from a non-farming family background. The system designed to support new farmers from such backgrounds generally requires applicants to demonstrate enthusiasm for all aspects of agriculture and rural life, and the system he applied to was no exception. It was likely during his training on the current farm that he developed the enthusiasm he expressed during the application process. The connection between employment support organizations and the new farmer support system is one of the key outcomes of this project. Nevertheless, it remains unclear why the experience of immersion and the subsequent physical strength gained should lead to social adaptation. While the AS1 members did not dismiss the value of the experience itself, they questioned whether it had been adequately evaluated.

What, then, is the intrinsic value of "immersion," which has been overlooked in the previous discussion? Kusayanagi (2004), referencing Goffman's "Theory of Order in Interaction," contends that "people cannot be fully immersed in a situation involving others face-to-face as long as they are making an effort to understand the situation and monitor how their body and behavior appear" or "as long as they are striving to perfectly adapt to the 'present order' of society and comply with it." Conversely, immersion occurs only when individuals are allowed to stop self-monitoring (Fig. 2.8). This aligns with the responses from SO2 instructors in interviews, where they noted that "there is nothing contrived in the farming environment,"

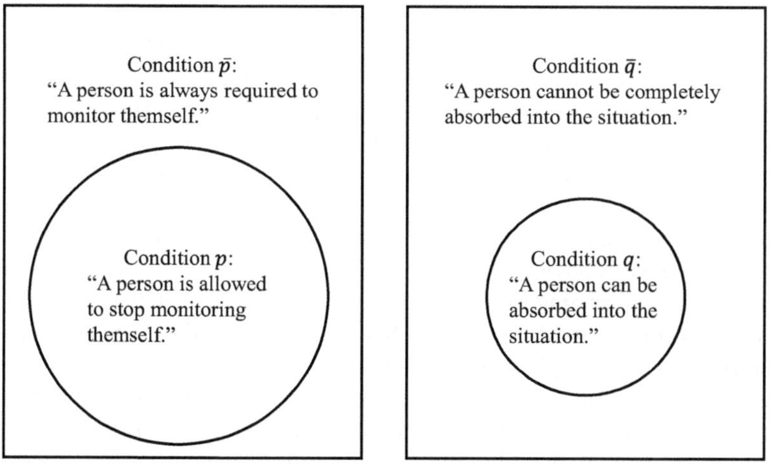

Fig. 2.8 Relationship between self-monitoring and absorption into a situation. According to the law of contraposition, namely, "If $\overline{p} \Rightarrow \overline{q}$ then $q \Rightarrow p$", the fact that a person can be absorbed into a situation implies that the person is allowed to stop monitoring themself

that trainees "were allowed to be who they truly were," and that "they did not need to be concerned with how they appeared to others."

The preceding discussion indicates that, during immersion, individuals should feel liberated from self-monitoring. This liberation occurs during farmwork, prior to the gain in physical strength. Two questions arise: first, why does farmwork produce a sense of liberation, and second, why does it lurk behind the physical benefits? The first question may be answered by the observation that the instructors guided their trainees to the farm with the intent of fostering this experience. However, this seems a matter of course and warrants little further elaboration. The critical issue is why these instructors specifically sought out farmwork opportunities for their trainees. In other words, why did they expect that only farmwork could provide this liberating experience?

Before addressing this, it is necessary to clarify what "liberation" from self-monitoring entails. From the perspective of disability studies, Hoshika (2007) argues that "interaction situations" involve power dynamics, with individuals divided into superior and inferior groups. The "superior" group holds the power to arbitrarily manipulate the "present order" at will. Because of fear of abrupt changes in it, those in the inferior position are reluctant to stop monitoring themselves. They are captives who could potentially be liberated from the volatile control of the superior group. This dynamic applies to those marginalized by the "hyper meritocracy" in contemporary society, where human ability is increasingly used as a metric for evaluating worker performance and distributing social positions (Honda 2011). In this context, "human ability" often refers to social skills, which are vaguely defined to justify the marginalization of individuals who are perceived to lack these skills. It is from this inferior status that instructors aimed to free their trainees.

The next question concerns how farmwork experiences provide the "inferior" with a sense of liberation. To answer this, we must examine the intrinsic nature of farmwork by comparing it with other occupations, particularly those in urban settings. The urban working environment is characterized by "interaction situations" where social norms, such as "business manners," dominate the "present order." In stark contrast, farmwork immerses individuals in tasks governed by the natural world. For example, farmers must complete ridge-making swiftly when rain is forecast, or harvest leaf vegetables before they bolt (as noted in the following chapter). In these situations, personal appearance is irrelevant. The strict requirements of physiology and climatology dictate decisions, leaving little room for artificial social norms to influence behavior.

Participants in farming activities inevitably perceive that they are subject to forces that, according to Morita (1987), humans cannot control. The lawlike properties inherent in natural phenomena limit the "superior" from enforcing artificial norms against the "inferior." Even when the "superior" provides directions to the "inferior," these must align with the laws of nature and be reasonable to all. In this context, participants realize that the "present order," which they have been compelled to maintain, is the result of human arbitrariness. Consequently, the "present order" is relativized, if not neutralized. This realization provides relief to those who have struggled with "interaction situations." Employment support service users

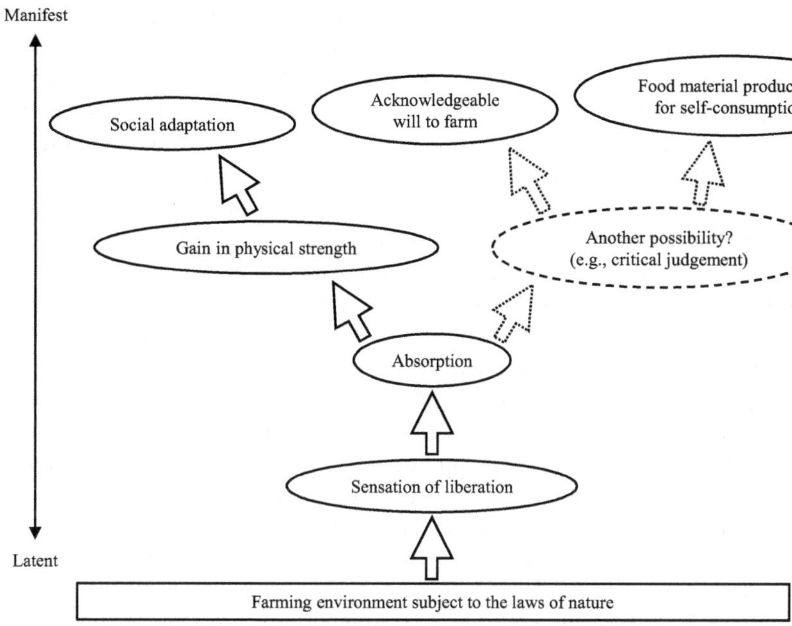

Fig. 2.9 Line of logic that links the farming environment with occurrences in farming activities such as "the sensation of liberation," "absorption," and "gain in physical strength." Solid-white arrows indicate a cause-and effect-relationship between the two connected occurrences; the initial and final points show the cause and effect, respectively

should not be an exception. Farmwork experiences may have revitalized them, enabling them to reach their physical potential and persist with laborious tasks. Indeed, their improved physical strength is a by-product of this sense of liberation.

This discussion outlines the logical connection between "the sensation of liberation," "absorption," and "gain in physical strength" (Fig. 2.9). Among these, the latter two are observable phenomena, confirmed by the SO1 instructor and trainees, while the former is an unobservable substance underlying these phenomena. It is now appropriate to revisit a previously unanswered question: Why did the sensation of liberation remain less recognized than the gain in physical strength? A straightforward answer could be that phenomena generally eclipse the underlying substance. However, this self-evident answer offers little insight. A more rhetorical interpretation could suggest that the sensation of liberation could have developed in a direction other than the gain in physical strength, perhaps enabling the "inferior" to exercise critical judgment. The question remains: why did instructors and trainees consider only the latter as a product of the former? The answer may lie in their uncritical focus on occurrences relevant solely to social adaptation.

Within the framework of employment assistance aimed at social adaptation, the trainees in SO1 and SO2 understood that, given their "inferior" status, they were expected—if not explicitly urged—to adapt to the "present order," which included adhering to "business manners." In interviews, instructors acknowledged the importance of these manners, recognizing them as another form of social or cultural norm. However, they also understood that some trainees might resist these norms. Despite this, instructors insisted that a lack of knowledge about these manners—however superficial or trivial—would disadvantage trainees in society. From a utilitarian perspective, it may seem more reasonable for job seekers or trainees to merely pretend to comply with the "present order." This false obedience aligns with what the SO1 instructor attempted to teach by emphasizing "how one appears to others." Trainees might find such obedience thrilling, although they would likely feel more liberated if they were allowed to criticize the norms they find absurd. Insofar as the training aims at social adaptation, farmwork inevitably leads to either apathy toward absorption or a backlash against the sensation of liberation.

As long as instructors maintain a detached attitude, trainees are likely to undervalue this experience or dismiss it entirely. Consequently, the product of absorption becomes an overemphasis on the superficial result—the gain in physical strength. Does this justify the use of farmwork opportunities or the exclusion of participants who wish only to produce food for themselves? These are essential questions posed by AS1 members. Instead of focusing on social adaptation through physical strength, they value the act of growing crops, which allows them to produce their own food. However, this does not necessarily make growing crops more realistic than adapting to the labor market. It remains possible that growing crops is more difficult than adapting to existing labor structures. Besides, it is nearly impossible to empirically compare these two choices. More importantly, whether growing crops offers a greater sense of liberation than social adaptation remains an open question. As a result, if the director fails to effectively communicate the concept of growing crops, trainees and participants may once again miss valuable opportunities to experience liberation. This issue will be addressed further in the next chapter.

2.6 Conclusion

Farmwork opportunities have come to embody an antinomy—a contradiction between creating "social adaptation" foundations for jobless youth and critically examining the positive orientation toward such opportunities. This antinomy allows those involved in employment assistance to reflect on the process of adjusting clients to a labor market that, at one point, had excluded them. In contrast to current arguments, some research on collaboration between agriculture and social welfare has focused on the roles and effects of farming activities in promoting a "return to normal social life," or repeated efforts at social adaptation (cf. Nakamoto and Hu, Nakamoto and Hu 2015, Nakamoto and Hu 2016; Nakamoto et al. 2018).

This raises new questions: Should this antinomy be resolved? If so, what immediate steps can be taken? It is important to note that new job opportunities can emerge from the skills that homeless elderly individuals demonstrate in guiding younger generations. Their humble awareness of having failed in "social adaptation" plays a key role. These findings align with my previous personal experiences. Before the present action research began, I often observed that many homeless individuals, despite surviving social exclusion, possessed skills unrelated to "human ability" and thus deemed unsuitable for "social adaptation." However, I also discovered that these skills could be passed on to younger individuals if they gathered at a specific workplace. The priority should be to create such workplaces, allowing the homeless participants' skills to evolve into new vocations. In this context, they would no longer be clients of support organizations, but rather, key contributors to job creation.

The next chapter addresses the skills required for growing crops and demonstrates that these are entirely different from the skills maintained by the homeless elderly. While this may initially suggest to readers that farming is not an appropriate job for social inclusion, this argument will be countered in subsequent chapters.

References[3]

***Agricultural Administration Office of Environment, Agriculture and Fisheries, Osaka Prefecture [Ōsakafukankyōnōrinsuisanbunōsēshitsu] (2007) A complete handbook for would-be new farmers [Shūnō Marugoto Handobukku]

*Akiyoshi Y (2011) A reformation of the agricultural system of Japan with reference to international case studies. Macro Rev 23(2):41–48

Dickens L, Watkins K (1999) Action research: rethinking Lewin. Manag Learn 30(2):127–140

*Hashimoto S (2007) A study on issue of community business. Bull Faculty Sociol Kansai Univ 38(2):5–42

***Honda Y (2011) The straining society: education, employment and young folks in the present situation [Kishimu Shakai: Kyōiku, Shigoto, Wakamono no Genzai]. Kawade Shobo Shinsha, Tokyo

***Hoshika R (2007) What is handicap: towards a theory of disability [Shōgai to ha Nani ka: Disuabiriti no Shakairironn ni Mukete]. Seikatsushoin, Tokyo

***Investigation and Research Section of Osaka Personal Support Scheme [Ōsaka Pāsonaru-sapōto Jigyōtyōsakenkyūbukai] (2012) Interim summary of the Osaka personal support project [Ōsaka Pāsonaru-Sapōto-Moderupurojekuto Jigyō Tyūkan Matome]

***Investigation Committee for National Survey on Homelessness [Hōmuresu no Jittai ni Kansuru Zenkokuchōsakentōkai] (2012) An annual report [Hōkokusho]

***Itō M, Kimichi M (2004) Developments in the policy for self-reliance and employment support to the youth and its challenges for the future [Wakamono no Shūgyō-Jiritsu o Shien Suru Sēsaku no Tenkai to Kongo no Tenkai: Mugyōsha ni Taisuru Taiō o Tyūshin to Shite]. Mon J Jpn Inst Labour 533:49–55

**Iwata M (2008) Social exclusion: lack of participation, uncertain belongingness. Yuikaku, Tokyo

***Japanese Association of Certified Social Workers (2010) A practical handbook on employment assistance from the viewpoint of social work [Sōsharuwāku Siten ni Motozuku Shūrōshien Jissen Handobukku]. Chuohoki Publishing, Tokyo

*Kanai K, Shikata M (2013) Analysis of employment support program for welfare recipients. Soc Policy Labor Stud 5(2):87–100

***Kashiwara City Social Welfare Council and Kashiwara City (2004) A tentative plan for revitalization of Johtokudani Valley as Citizens' Home (Provisional name) [(Kashō) Jōtokudani Shimin Furusato Zukuri Kōsō (An)], Jul 2004

[3] *: in Japanese with an English title and summary, **: in Japanese with an English title, ***: in Japanese.

***Kodera M (1986) Research on changes in grape cultivation of Osaka Prefecture [Ōsakafu ni okeru Budōsaibai no Rekishitekihensen ni Kansuru Kenkyū]. Doctoral dissertation, Osaka Prefectural University

***Kondō T (ed) (2013) Farming by people with disabilities through collaboration between the agriculture and welfare sectors [Nōfukurenkē ni yoru Shōgaishashūnō]. Soshinsha, Tokyo

**Kudo K (2013) How to nurture the careers of personnel helping underserved youth. Mon J Jpn Inst Labour 639:63–71

***Kusayanagi C (2004) "Ambiguous predicaments" and society: the sociology of claim making ["Aimaina Ikizurasa" to Shakai: Kureim Mōshitate no Shakaigaku]. Sekaishisosha, Kyoto

*Matsumiya A (2013) Developments in quasi-farming activities led by non-farmers in urban areas: a case study in Nagakute City and Nisshin City of Aichi prefecture. In: Sustainability research. Institute for Sustainability Research, Hosei University, pp 85–97

***Ministry of Agriculture, Forestry and Fishery (2007) Stabilization of farm management through long-term farmland use, vitalization of rural communities through facilitation of new entry into agriculture by Retirees and Suchlike, and contemporary appropriateness and necessities of measures in accordance with the idea centered upon land-owning farmers [Nōchi no Tyōkitekiriyō ni yoru Ninaite no Nōgyōkēē no Antēka, Tēnenshunōtō Shinkisannyū no Sokushin ni yoru Chīki no Kasseika, Jisakunōshugi ni Motozuku Sochitō no Konnishitekidatōsē], May 2007

***Ministry of Agriculture, Forestry and Fishery (2010) To protect the precious farmland: how to promote the countermeasure against its abandonment [Kakegae no Nai Nōchi o Mamoru Tame ni: Kōsakuhōkichitaisakusuishin no Tebiki]

*Mishiro Y (2012) Difficulties faced by Agricultural College for playing their role in increasing new farmers in Japan. Kumamoto Univ Stud Soc Cult Sci 10:241–263

***Morita S (1987) Agriculture is agriculture: a trick of the modernization theory [Nōgyō ha Nōgyō dearu: Kindaikaron no BŌRYAKU]. Rural Culture Association Japan, Tokyo

*Nakagawa S, Tetsuo I, Yukihiro I, Miki K (2006) The strategies for the revitalization of local agricultural areas: a plan of Agricultural Business School. Bull Tottori Univ Environ Stud 4:53–74

*Nakamoto E, Hu B (2015) The present state and issues of agricultural activities in supporting youth employment. J Rural Probl 51(2):116–121

*Nakamoto E, Hu B (2016) The roles of agricultural and horticultural activities in supporting *hikikomori*'s (people who withdraw from society) return to a normal social life: an empirical study on the medical and welfare effects of agriculture through an experimental social scientific approach. J Rural Econ 87(4):319–333

*Nakamoto E, Yamamoto K, Hu B (2018) Examination on mood changes and the medical/welfare effects obtained from agricultural activities: a case study of NEET aiming return to normal social life. J Rural Econ 90(1):17–22

***Nakayama T, Hashimoto S (2006) New job creation and community revitalization [Atarashī Shigotozukuri to Chīkisaisē]. Bunnrikaku, Tokyo

**Ogata K-i (2013) The strategy for reducing abandoned farmland in the Agricultural Land Act revision in 2009 and its application in present situation. Kochi Univ Rev Soc Sci 106:75–103

***Okamoto Y (2005) On the plan for revitalization of Johtokudani Valley as citizens' home (provisional name) [(Kashō) Jōtokudani Shiminfurusatozukurikōsō nit suite], 18 Apr 2005

***Ōsawa S (2010) Construction of an employment facilitation system in the field of agriculture for people with intellectual disabilities and its implementation [Nōgyōbun-ya ni Okeru Chitekishōgaisha no Koyōsokushin shisutemu no Kōshiku to Jissen]. Mirai, Gifu

**Satō Y (2008) Looking at the loss of the desire to communicate in tandem with support for youth independence: a paper written in response to the survey report dealing with young people in the 'neet' state. J Tokyo Keizai Univ 258:71–85

Susman GI, Evered RD (1978) An assessment of the scientific merits of action research. Adm Sci Q 23(4):582–603

*Takebe T (2011) An examination of the possible problems of the entry by stock companies into agriculture by leasing farmland. Nat Resour Econ Rev Kyoto Univ 16:1–16

***Tanaka M, Toshikatsu D, Yoshihumi H, Hiroyuki U, Yoshinori Y, Takaaki I (2011) Finding employment in the field of agriculture with disabilities: through practice in a firm [Nōgyōbun-ya ni Okeru Shōgaishashūruō: Jigyōshogemba no Jissen o Tōshite]. Res Bull Shujitsu Univ Shujitsu Junior Coll 40:61–72

***Toyohara N, Yōchi I, Kae M (2007) Agricultural and horticultural activities for well-being: a manual of how to practice effortlessly [Fukushi no Tame no Nōengēkatsudō: Muri Sezu Dekiru Jissen Manyuaru]. Rural Culture Association Japan, Tokyo

*Tsunashima H (2015) Significance of job opportunity creation through a farmland recovery project for people with difficulty in obtaining employment. Food Agric Environ 16:99–114

***Tsuno Y (1995) Small agriculture: a quest from mountainous farm villages [Chīsai Nōgyō: Sankanshinōson kara no Tankyū]. Rural Culture Association Japan, Tokyo

***Ushino T, Yuko N, Kenishi H (2007) A study about employment of people with intellectual disabilities in agriculture: manpower for agriculture and employment for the disabled [Nōgyō ni Okeru Chitekishōgaishakoyō ni Kansuru Ichikōsatsu: Nōgyō ni Rōdōryoku o, Shōgaisha ni Koyō no Ba o]. J Rural Plan 25(4):556–563

*Yasunaka S, Hitoshi Y, Chie K, Kenji I (2009) Identifying problems in and suggesting solutions for the provision of support by classifying work style of persons with disabilities in agriculture. Tech Rep Natl Inst Rural Eng 210:49–59

Chapter 3
Analysis of the Human-Plant Relationship in Agriculture-Related Job Opportunity Creation

Abstract If directors fail to adequately communicate the essence of crop cultivation, trainees or participants may miss valuable opportunities to experience a sense of liberation. To address this, I aimed to provide a comprehensive perspective on the challenges they would face while acquiring farming skills. Using the Modified Grounded Theory Approach, I analyzed 232 cases and identified 37 concepts, which were grouped into six categories. In this context, "operation" refers to a component of the cropping system, defined in terms of the physiological needs of crops, while "motion" represents the human physical actions involved in these operations. Metacognitive awareness of "cognitive characteristics" and supplementary "prerequisite knowledge" are essential in forming the mental foundation for "motion." The mental activities that precede "operation" include "observation" and "decision-making." Many participants successfully overcame the challenge of "operation," yet hesitated to "observe" various changes in the field. This sequence of challenges represents a two-step transformation in the relationship between participants, their director, and the crops. At present, the key issue is how to transition into the second phase, where a qualitative change in spontaneity occurs. The reasons for this challenge are examined through interviews with participants who were not included in the previous chapter.

Keywords Act of growing crops · Autonomous farming activities · Employment assistance · Instruction giver-receiver relationship · Spontaneity

3.1 Introduction

The previous chapter outlined several benefits of incorporating farmwork into employment assistance programs for young trainees. This approach may not only foster job creation for elderly homeless individuals, allowing them to demonstrate their abilities, but also offer an alternative to traditional employment assistance,

The original text written in Japanese, Tsunashima (2018), underwent major revision for Sects. 3.1–3.3 and 3.4.1 of the present chapter.

H. Tsunashima, *Farming and Social Inclusion in Japan*, Urban Agriculture,
https://doi.org/10.1007/978-3-032-01598-3_3

which typically focuses on individual social adaptation. However, it is important to recognize that these findings were based on interview results, which warrant critical scrutiny. Since most interviewees had no prior experience in farming, there is a considerable possibility that their perceptions of agriculture were flawed.

As mentioned in the previous chapter, participants, some of whom were interviewees, lacked the necessary experience and thus required guidance. What sets farmwork apart from other occupations is the rationality of the instructions, grounded in the natural laws of agriculture. As long as these instructions are based on natural principles, the relationship between the giver and receiver of instructions should be less hierarchical than the typical superior-subordinate dynamic observed in urban employment settings. While acknowledging the potential benefits of the "instruction giver-receiver relationship" in farming, the existence of this relationship itself requires further examination. This is the focus of the present chapter.

A key question is whether such relationships could lead to significant issues. Participants often sought guidance on every farm task, although alternative methods for addressing their knowledge gaps could have been pursued. For instance, an inexperienced gardener or farmer wishing to grow their own food would typically seek out information to overcome their lack of knowledge. In contrast, job seekers and trainees often expect assistance to bridge their knowledge gaps. Practitioners are compelled to support these individuals within the allotted time unless willing to perform the tasks on their behalf. Time constraints become particularly acute when practitioners are working to secure funds for labor wages through farm production. Thus, the "instruction giver-receiver relationship" emerges more from the attempt to create jobs and provide employment assistance than from the inherent nature of farming activities. In hindsight, members of the alternative school, who might have shared the wish to learn agriculture as described in the previous chapter, failed to critically assess this relationship, to which they too became susceptible.

What problems arise when participants view the request for instructions on crop cultivation as a right? From the opposite perspective, this relationship assumes that there will always be someone prepared to provide guidance. However, this assumption is misguided, as it is unlikely that anyone can possess complete knowledge in advance. What if the potential instructor is also uncertain about how to address an unforeseen issue? This expectation raises several questions, including whether job creation initiatives should assume the existence of an omniscient instructor when instruction should cease, and the extent to which instruction should be detailed. Ultimately, what does it mean to solicit, issue, and follow instructions on crop cultivation? Addressing these issues is key to correcting misconceptions about agriculture that may have been shared by participants.

In parallel with my growing skepticism toward the "instruction giver-receiver relationship," a national survey of "Certified Farmers"[1] in Japan was conducted on

[1] The farmers whose farm management improvement plans, based on their original ideas and aligned with the foundational reinforcement concepts introduced by their municipal government, have been approved by the local or central government, thereby qualifying them for substantial public support.

the state of the "partnership between agriculture and social welfare." The survey found that 62% of respondents were still reluctant to participate in this partnership, and 12% had already discontinued their involvement. In this context, "participating in the partnership" generally refers to farmers employing clients from social welfare organizations who possess skills, if any, outside of farming. Two primary reasons for reluctance were: (1) the laborers were only capable of performing unskilled tasks, and (2) providing instructions to laborers became an undue burden for the person in charge (Kotani et al. 2016). These issues reflect farmers' disappointment with the low skill levels of their laborers and the additional burden of instructing them. Farmers faced a difficult choice: either limit laborers to tasks that required no special expertise or dedicate time and resources to assisting them with more complex tasks. Together, these two points underscore the role of the "instruction giver-receiver relationship" in the agriculture-welfare partnership, which mirrors the dynamic in employment and job training. In short, farmers engaged in this relationship for the same reason as in the present farm program.

Clearly, practitioners struggle to manage unskilled laborers with limited agricultural experience. Some may attempt to improve their laborers' skills but often find this to be an arduous task. However, once laborers acquire farming skills independently, the burden on practitioners is alleviated, leading to increased labor productivity and potentially more job opportunities. More importantly, increased autonomy likely enhances the mental well-being of laborers, as I have observed firsthand. For instance, if a laborer develops a deep understanding of farm tasks and the rationale behind them, the practitioner may allow them to make decisions independently, resulting in a sense of accomplishment and increased motivation. This process is akin to the "aha" moment in psychology (Shen et al. 2016) and contrasts with the "sense of liberation" described in the previous chapter, which requires only minimal engagement from the laborer. Several studies on horticultural therapy and care/social farming have conflated these two concepts (cf. Elings 2006; Haller 2017). Improving the instruction giver-receiver relationship and enhancing laborers' skills will create a mutually beneficial relationship between practitioners and laborers. Therefore, addressing this relationship—whether to resolve or reform it—has been a critical issue for practitioners. Yet, it has been largely overlooked in academic inquiries into urban agriculture, both in Japan and the West.

Past research on the agriculture-welfare partnership in Japan has largely assumed the inevitability and unquestionability of the relationship between directors and laborers. Several studies, including Kotani et al. (2016), suggest that this asymmetric relationship stems from the gap between experienced and inexperienced individuals. However, these studies treat the relationship as a fixed and unchangeable condition, thereby closing off avenues for potential autonomous activities (e.g., Katakura et al. 2007; Nakamoto and Sawano 2020; Iiba and Yamabata 2022). Similarly, research on care/social farming in Europe shows little inclination to promote the development of laborers' skills and autonomy. A few studies briefly mention farming skills, but they tend to focus on behavioral improvements in individuals with mental disabilities or disorders during farming activities. Notably, Moruzzo et al. (2019) and Torquatti et al. (2019) assess clients' skill levels only to determine task compatibility, without examining how skill acquisition may contribute to overall well-being.

In contrast, past research on urban farming and gardening in the West does not assume a hierarchical relationship between growers and more experienced individuals. Some studies depict urban farmers as largely independent, often without technical guidance. For example, senior citizens are described as returning to their rural roots in urban allotment gardens (Tei and Gianquinto 2010, as quoted and translated in Gasperi 2015), and Mexican immigrants in the U.S. can create "can grow" lists for crops (Bellows et al. 2010). Additionally, Cabannes and Raposo (2013) and Opitz et al. (2016) note that some urban dwellers acquire gardening skills through experimentation and experience, without formal agricultural education. However, these studies do not address the factors that enable such autonomous skill development.

In summary, existing research lacks a comprehensive perspective on the challenges associated with instruction giver-receiver relationships. Specifically, there is insufficient exploration of the obstacles faced by laborers without farming experience or jobless youth without experience in any field when they attempt agricultural activities and learn the necessary skills. This chapter, therefore, examines these obstacles in the context of skill acquisition, drawing on the continuation of the action research project discussed in the previous chapter. Additionally, it addresses a question hidden in the previous chapter regarding why homeless elderly laborers did not experience the same sense of liberation as trainees receiving employment assistance.

3.2 Study Methods

3.2.1 Data Collection

As discussed in the previous chapter, I served as the director, providing instructions to participants, and encountered several challenges. Communication of farm tasks often relies on what is referred to as "tacit knowledge" in management studies (cf. McAdam et al. 2007). This type of knowledge makes it unrealistically time-consuming to fully verbalize the intended instructions. For example, when participants with no prior farming experience were instructed to "sow this ridge with radish," they struggled to understand what to do unless they knew the growing conditions for radishes through to harvest. In some cases, they required detailed supplementary instructions. For instance, I had to explain that they should first dig holes to a specific depth and place the seeds at a particular interval (Fig. 3.1). This was a time-consuming process.

In such circumstances, I needed to issue instructions efficiently to ensure participants could complete tasks within the time constraints. I observed that more experienced participants required fewer detailed instructions. Consequently, I gradually

Fig. 3.1 Example of ridge preparation before sowing radish. After the director adjusted the ridge height with a hoe, a participant dug 2 cm-deep holes at a 30 cm-interval to bury radish seeds along the centerline of the ridge

shortened the instructions as participants gained experience. However, this sometimes led to misunderstandings, and I realized in hindsight that these "happenings" were often linked to overly brief instructions. What was omitted in these condensed instructions corresponded to key learning moments during skill acquisition. I documented these incidents in field notes with photographs and reviewed them later to identify what additional information I could have provided to supplement the instructions. In some cases, I also noted whether the supplementary instructions were successful.

3.2.2 *Data Analysis*

I employed the Modified Grounded Theory Approach (M-GTA) to analyze these records. The analysis focused on "changes in things to be learned during the increase in autonomy." In this context, "autonomy" refers to the ability to perform a farm task without detailed instructions. In other words, as participants required less detailed or more general instructions, their autonomy increased. The analysis centered on participants from the farm, as described in the previous chapter, including homeless laborers and young trainees receiving employment assistance. Additionally,

I documented the speech and behavior of instructors involved in communication between the director and participants when they played an active role.

In the first step of the analysis, I entered the descriptions of "happenings" into an "Analysis Worksheet." Events before March 2015 were categorized under "open coding" to generate "concepts." Subsequently, "happenings" from April 2015 to March 2017 were used for "selective coding" to elaborate on the concepts. During this process, provisional concepts were continually reviewed. Once these concepts were classified based on their similarities, the selective coding phase concluded, and the analytical process reached "theoretical saturation." Finally, I described the interrelationships among the categories and presented them in a "storyline."

It is important to note that M-GTA differs from other grounded theory approaches, such as those described in publications by Jackson et al. (2017) and Fendel et al. (2018), which present variations of the grounded theory methodology. M-GTA, originally proposed by the Japanese sociologist Yasuhito Kinoshita (1999, 2003, 2005, 2007), has been primarily disseminated in Japanese. However, Kambaru (2018) introduced it to the international community in English. While M-GTA is not widely known globally, increasing research based on it is being published in English (e.g., Nagayama and Hasegawa 2014; Okura 2019).

Like conventional grounded theory approaches, M-GTA assumes that the data to be analyzed are verbal, typically collected through interviews. In this study, however, the primary data came from my observations, supplemented by occasional verbal records of conversations with participants. Some readers may question how the objectivity of these data is ensured. In response, Kinoshita (1999) argues that the success or failure of M-GTA should be judged by the extent to which its results are useful to practitioners. From this perspective, the application of M-GTA in this study has proven appropriate, as it has effectively guided subsequent investigations discussed in later chapters.

3.2.3 Additional Interviews

Starting in fiscal year 2014, "The Organization to Support the Homeless in Kamagasaki"[2] began recruiting homeless individuals and welfare recipients from Nishinari Ward, Osaka, to work on the farm. The Nishinari Ward Social Welfare Council also helped to find outlets for the farm's produce. By the end of March 2017, the total number of participant workdays amounted to 3655. Among the participants, elderly welfare recipients worked without pay, while homeless laborers and younger welfare recipients received compensation. The daily wage of JPY 5700 was based on the Special Scheme of Job Opportunity Provision for the Elderly (see Chap. 2), which the organization managed.

[2] The organization, founded in 1999, aims to enhance social welfare through activities that improve the social treatment of the homeless and those at risk while establishing a local community to support their independent livelihoods (http://www.npokama.org/index.html).

Fig. 3.2 Interview with a participant by volunteers from a university

To assess the effectiveness of the job creation initiative, I conducted semi-structured interviews with 17 new participants between January and March 2016. These included two younger welfare recipients, seven homeless laborers selected by lottery, and eight individuals who expressed interest in continued employment on the farm. The interviewees were asked the same questions as the participants mentioned in the previous chapter. To facilitate candid responses, university student volunteers conducted the interviews on my behalf in nine separate sessions (Fig. 3.2). This chapter presents only a partial disclosure of the interview results to inform the M-GTA analysis.

3.3 Results of the Analysis Through M-GTA

3.3.1 Storyline

An analysis of 232 cases identified 37 "concepts," which were organized into six "categories" (Table 3.1). The resulting "storyline" illustrates the process of crop cultivation: Defined by the physiological needs of crops, the term "operation" refers to a component of a cropping system and can be broken down based on human physical actions. Each part of the "operation" corresponds to a specific "motion." For example, the "operation" of seeding includes multiple "motions," such as making a hole, placing the seeds, covering them with soil, and watering. Metacognitive

Table 3.1 Categories and concepts represent things to be learned during the increase in autonomy

Category	Subcategory	Concept (no.)
Motion		How to hold and move farming tools (01)
		How to hold a body of crop plants (02)
		How to carry goods along the shortest possible path (03)
		How to increase work accuracy without sacrificing time efficiency (04)
		How to increase time efficiency when allowed to be inaccurate (05)
Prerequisite knowledge	Knowledge exclusively used in the farming industry	Knowledge of technical terminology (06)
		Knowledge of plant physiology (07)
	Commonly applied knowledge	Knowledge of common place motion (08)
		Knowledge of commonplace natural phenomena (09)
		Knowledge of quantitative expressions (10)
Cognitive characteristics	Inconsistencies additional verbal instructions can resolve	Consciousness of the objective of the current "operation" (11)
		Balance of concentration across multiple objects (12)
	Inconsistencies require additional nonverbal instructions	Spatial cognition (13)
		Distinction among plant spices by visual and tactile characteristics (14)
		Fine motor skills needed for precision tasks (15)
Operation	Integration of motions	Proper arrangement of "operational" procedure (16)
		Proper choice of tools (17)
		Awareness of difference in conditions (18)
		Interpretations of instructions with appropriate standards (19)
	Wider perspective on "operations"	To make the next "operation" efficient (20)
		To anticipate what will happen in the natural world and take precautions against it (21)
		Consideration for coworkers (22)
		Consideration for consumers (23)

Observation	Review of "operations" done in the past	To check the results of "operations" done in the past (24)
		To monitor the plant's progress after "operations" (25)
		Comparisons between the current state of the observation object and its previous condition (26)
		To confirm the results of their director's trial (27)
		To deduce general principles from previous instructions on various "operations" (28)
	Utilization of opportunities to see things ordinarily invisible	To examine the underground situation (29)
		To examine the interior and taste of crops (30)
	Utilization of unconcerned opportunities	Unconscious observation (31)
		Confidence in being more informed than the director (32)
		Habitual observation of the entire farm (33)
Decision making		To determine whether a particular "operation" should be performed at a specific time (34)
		To stop expecting instructions and observe the crops instead (35)
		To prioritize a limited number of "operations" based on "observation" results (36)
		To learn from failures (37)

awareness of "cognitive characteristics" and supplementary "prerequisite knowledge" are important for forming the mental foundation necessary for executing these "motions." Mental activities that precede the "operation" include "observation" and "decision-making."

As a director, I had to perform both actions before providing instructions to participants. The former involves gathering information about the crops and environment, which is essential for the latter—deciding what the next "operation" should be. However, "observation" also serves to assess the success or failure of previous "operations." In essence, "operation" follows decision-making, yet it also requires subsequent observation. These three actions—observation, decision-making, and operation—form a cycle that repeats. The relationships among the six "categories" demonstrate the central theme of the analysis: "changes in what is learned as autonomy increases" (Fig. 3.3).

Initially, a complete beginner must follow instructions to complete an "operation," which is divided into several "motions." At this stage, the director must explain each step, corresponding to a specific "motion." Some beginners quickly learn to complete certain "operations," while others struggle. This suggests that some beginners require additional guidance on how to execute appropriate "motions." When these supplementary instructions are insufficient, the director may need to address the participant's "cognitive characteristics" and fill any gaps in their "prerequisite knowledge." However, there are instances where no adequate countermeasures can be taken, especially when the "motion" requires physical skill.

The preliminary goal for beginners is to perform basic farm tasks using simple instructions. Once this goal is achieved, the next step is for participants to understand the meaning behind the sequence of "motions" in an "operation" and grasp its essence. This is why the current action research project was designed to have

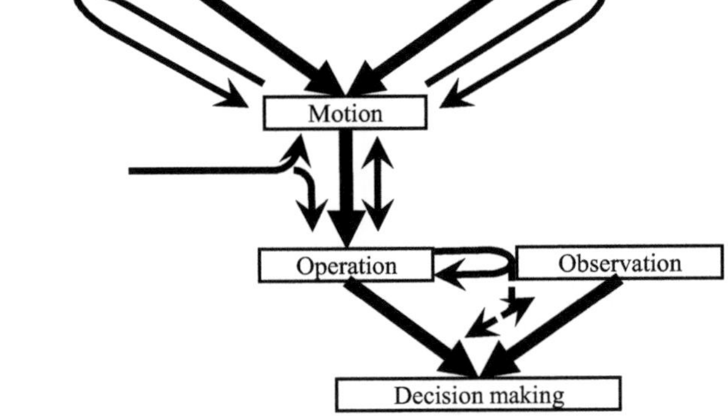

Fig. 3.3 Things to be learned during the increase in autonomy. Bold arrows indicate the direction to which the autonomy enhanced, while the thin and broken ones are the routes actually observed and those yet to be actualized, respectively

participants repeat the same "operation" under different conditions, such as unfamiliar crops or environments. As a result, some participants successfully learned several "operations."

After this achievement, the next goal was for participants to engage in "decision-making" that initiates an "operation" without prompting. In the context of this study, this meant participants identifying the farm tasks they needed to perform on specific days. It was crucial for them to determine the proper timing for each "operation," which required them to "observe" the crops and environment in the field. However, most participants hesitated to make decisions in practice, often avoiding "observation" of the various changes in the field that would inform the need for a given "operation." The process by which learning "operations" leads to "observation" and "decision-making" remains unclear.

3.3.2 *Concepts and Subcategories as Components of Categories*

3.3.2.1 Components of the Category "Motion" (Five Concepts)

These concepts represent the perspectives from which each "operation" is divided into several "motions." Initially, basic complementary instructions were provided from three primary viewpoints: "how to hold and move farming tools" (Concept-01), focusing on energy conservation while considering the specific purpose for which each tool is designed; "how to hold a body of crop plants" (Concept-02), ensuring proper planting density and maintaining product quality up to the moment of sale; and "how to carry goods along the shortest possible path" (Concept-03), aiming to save energy through strategic placement of goods around participants. In this context, "goods" typically refer to tools and objects, which are either handled by the tools or directly by the participants' hands.

Further instructions were provided to improve work efficiency: "how to increase work accuracy without sacrificing time efficiency" (Concept-04), exemplified by the creation of new tools using readily available materials; and "how to increase time efficiency when allowed to be inaccurate" (Concept-05), or using a more casual approach, allowing participants to save time and energy when possible.

In some instances, verbal instructions, such as "when weeding crop-free sections, pull the hoe toward you to uproot weeds while stepping forward incrementally," successfully prompted participants to perform the appropriate motions. However, other situations required additional nonverbal interventions, such as model demonstrations or physical assistance from the director. For example, after demonstrating ridge making—which involves leveling the ground and ridge surfaces with a hoe—the director instructed a participant to hold the hoe's shaft in a specific posture so they could better understand how to move the blade horizontally. Occasionally, these interventions were not effective.

3.3.2.2 Components of the Category "Prerequisite Knowledge Supplements" (Two Subcategories Incl. Five Concepts)

This category consists of two subcategories, each representing intellectual foundations necessary for understanding instructions related to "motions." When participants lacked this knowledge, the director was required to provide supplementary information.

Subcategory the "Knowledge Exclusively Used in the Farming Industry"

This subcategory consists of two key concepts. First, "knowledge of technical terminology" (Concept-06) includes terms such as *family*, *variety*, and *bolting*, which participants typically do not understand without practical farm experience. Second, "knowledge of plant physiology" (Concept-07) is essential for effective weeding and watering. Weeding, for example, requires understanding how weeds propagate and identifying parts of plants that are likely to germinate or sprout, which should be removed. Some species propagate from stem nodes or root fragments, as is often the case with Poaceae and Asteraceae. Thus, understanding terms such as *family* and distinguishing stems and roots is critical. This knowledge helps participants understand the reasons behind various farm tasks.

In this context, sweet potato served as a useful teaching example. The crop has an underground storage organ, similar to the potato. However, while the sweet potato is a root, the potato is a subterranean stem. Consequently, sweet potatoes grow downward, whereas potato stolons spread horizontally, which requires different ridge preparation methods. This distinction would suggest that these plants belong to different families. Participants were surprised to learn that, unlike potatoes, which belong to the Solanaceae family, sweet potatoes are part of the Convolvulaceae family, sharing this classification with morning glory (*Ipomoea nil*) and Chinese water spinach (*Ipomoea aquatica*), both familiar to the Japanese as ornamental plants and summer vegetables, respectively. These plants are also affected by the same pest, a species of gelechiid moth (*Helcystogramma triannulellum*). This lesson sparked participants' interest in the relationship between the term *family* and the organ that constitutes the edible part of the crop.

Regarding the task of watering, I often found that I had failed to explain its varying purposes depending on the crop's growth stage. Watering is not merely for hydration; it facilitates germination, enhances vegetative growth, and preserves fruit freshness. Consequently, participants tended to indiscriminately water the soil around the plants without adjusting the amount of water based on the crop's needs. The difficulty lies in the time delay between the watering and its effects on the plants, making it hard to assess whether the task was done correctly. Participants can only gauge success once they begin to consciously observe the results, as outlined below. Therefore, the director must impart this type of knowledge to participants whenever possible, unlike the knowledge in the next subcategory.

Subcategory "Commonly Applied Knowledge"

This subcategory includes three concepts that relate to forms of knowledge participants may have acquired through vocational experiences in industries outside of agriculture, such as construction or domestic work. The first, "knowledge of commonplace motion" (Concept-08), involves tasks such as tying a string and avoiding careless mistakes. For example, when harvesting fruit or vegetables, it is important to inspect each branch from multiple angles to prevent overlooking fruits in blind spots. The second, "knowledge of commonplace natural phenomena" (Concept-09), pertains to life and physical phenomena. For instance, it includes understanding how to distinguish male and female flowers, recognizing that steamy conditions encourage the growth of mold or moss, and knowing that the lever principle can save effort. The third, "knowledge of quantitative expressions" (Concept-10), involves visualizing quantities described in instructions, such as "eight-tenths" or "five centimeters."

3.3.2.3 Components of the Category "Cognitive Characteristics" (Two Subcategories Incl. Five Concepts)

To perform each "motion" correctly, participants must grasp the conditions of the object matter described in the instructions and the surrounding circumstances. However, participants sometimes display cognitive characteristics that deviate from the director's expectations. Thus, uniform instructions from the director may not always elicit the intended actions. The following concepts explore different perspectives from which the director and participants can collaborate to address inconsistencies between the intended actions and those actually performed. These efforts aim to foster metacognition of cognitive characteristics.

Subcategory "Inconsistencies Additional Verbal Instructions Can Resolve"

Participants may sometimes focus excessively on the specific "motion" at hand and lose sight of the broader objective of the "operation." In such cases, the director must offer additional verbal instructions to have a "consciousness of the objective of the current 'operation'" (Concept-11). For example, the "operation" of weeding involves a "motion" of cutting or pulling out weeds around crop plants. However, the ultimate focus should be on preserving the crop plants, while weeds can be roughly handled. Similarly, during the early stages in harvesting certain fruit vegetables, while fruits are the temporal focus of the action, the leaves and stems should be treated carefully to ensure the plants' vegetative growth. In instances where participants grapple with a set of objects scattered around, the director may need to offer verbal instructions on pertaining to the "balance of concentration across multiple objects" (Concept-12). This is particularly important for the participants with developmental disabilities who have difficulty maintaining overall attentiveness during even a "motion."

Subcategory "Inconsistencies Require Additional Nonverbal Instructions"

When participants struggle with "spatial cognition" (Concept-13), verbal instructions are often ineffective. For instance, if a participant is slow to level a ridge, providing specific instructions such as "plane off the soil here" or "fill the hollow with soil" may result in task completion. However, this does not enhance their spatial perception. In these situations, the director may need to visualize the gradients with water flow, if feasible.

Similarly, when a participant cannot draw a "distinction among plant species by visual or tactile characteristics" (Concept-14)—including colors, shapes, and textures—verbal instructions alone are insufficient. Recognizing differences between crops and weeds is essential for many "operations," requiring the director to devise nonverbal strategies to prevent confusion.

In some cases, participants may also lack the "fine motor skills needed for precision tasks" (Concept-15), such as handling delicate seeds or perishable leafy vegetables. In these instances, it may be impossible for the director to offer effective verbal guidance, and they may need to take over the task or ask another participant for assistance.

3.3.2.4 Components of the Category "Operation" (Two Subcategories Incl. Eight Concepts)

When the director aims to teach a participant to understand an "operation" as a series of integrated "motions," the participant must recognize the meaning of each individual "motion" and relate them to their past experiences. The following concepts outline the perspectives and goals that participants should develop at this stage.

Subcategory "Integration of Motions"

For participants who have completed the full process of a given "operation," I have established a rule: they must perform the entire "operation" without being prompted about its individual "motions." They are expected to independently assemble these "motions" into the complete "operation." This requires that participants remember the component "motions" of the current "operation" to independently make the "proper arrangement of the 'operational' procedure" (Concept-16) and the "proper choice of tools" (Concept-17). In this context, participants naturally rely on their prior instructions as a guide. However, these past instructions do not always align with the present situation. Therefore, participants may need to adjust what they recall in light of current conditions, such as weather, soil moisture, or crop species and variety. This necessitates the development of an "awareness of difference in conditions" (Concept-18) rather than a superficial application of past instructions.

Of course, directors must provide instructions in a specific manner among numberless possibilities. These instructions may have been effective in previous "operations" and could offer useful insights for future tasks, fostering greater autonomy among participants. However, they may also inadvertently mislead participants regarding the director's intentions, particularly when a similar "operation" is carried out, but details about its component "motions" are omitted. A recurring issue in the current farm context illustrates this point: during past seeding or manuring events, the director intended to demonstrate the appropriate density for sowing seeds or applying manure per unit of land area. However, participants were primarily influenced by the fact that the seed or manure package was emptied. As a result, participants sometimes used a full package of materials in subsequent tasks, regardless of the area. In such cases, the director must correct past instructions to clarify any misunderstandings. Specifically, the director should emphasize that it is the density of seeds or manure, not the total amount consumed, that affects crop growth.

In general, directors should encourage participants to provide "interpretations of instructions with appropriate standards" (Concept-19). If the director struggles to convey an image of the "appropriate standards" to a participant, they should avoid shortening the instructions. Doing so would suggest that the participant has yet to master the "operation."

Subcategory "Wider Perspective on 'Operations'"

Participants' perspectives when addressing individual "operations" are insufficient for effectively completing the entire crop production process. This process involves a series of interconnected operations, requiring close coordination across both time and space. Thus, adopting broader perspectives is essential.

First, "to make the next 'operation' efficient" (Concept-20) requires considering the temporal relationship between current and future "operations." For instance, participants should leave water spaces when filling seedling-raising pots and should weed or level the ridge while harvesting root crops.

Second, "to anticipate what will happen in the natural world and take precautions against it" (Concept-21) involves preparing for natural phenomena that may disrupt current operations. This is particularly important for participants receiving employment assistance, as they are constrained by a set departure time. These constraints may lead to additional tasks that would not be necessary if participants were able to finish their operations within the day. For example, if participants break ridges to apply basal fertilizer but must leave before completing the task, and they expect rainfall, they should repair the ridges by lifting dry soil onto them before the rain begins, to avoid handling heavy, soaked soil afterward.

Third, participants must also consider that the perspectives of others may differ. In tasks requiring collaboration, participants need to show "consideration for coworkers" (Concept-22). Furthermore, packaging and shipping farm produce require "consideration for consumers" (Concept-23), which entails understanding consumer preferences and behaviors regarding food consumption.

3.3.2.5 Components of the Category "Observation" (Three Subcategories Incl. Ten Concepts)

This section outlines the concepts guiding participants in effectively obtaining relevant information for decision-making regarding "operations." To be action-provoking as per the above "storyline," the "observation" process requires both an object of focus and a mental representation. Specifically, observation involves comparing the current state of a crop to its ideal condition, a comparison participants must visualize based on direct experience, secondhand information, or even supposition. If no disparity is identified, participants will be convinced that the latest "operations" were executed flawlessly. If a gap is found, it indicates an error or omission during the "operation," prompting participants to address the issue as soon as possible. These concepts are categorized into three subgroups based on the timing and method by which participants form an ideal comparison with actual crop conditions.

Subcategory "Review of 'Operations' Done in the Past"

When participants simply follow their director's instructions, they typically do not concern themselves with the outcomes of the "operations." However, for autonomous learners, it is crucial to reflect on both the successes and failures of past "operations." A key requirement is "to check the results of 'operations' done in the past" (Concept-24). For example, after sowing seeds, participants should verify whether germination has occurred the following day. Additionally, participants should not merely check the soil's moisture or the appearance of ridges after watering or weeding; they must focus on assessing the vitality and glossiness of the leaves and fruits "to monitor the plant's progress after operations" (Concept-25).

The next step involves "comparisons between the current state of the observation object and its previous condition" (Concept-26). When participants learn to make such comparisons, they can, for instance, deduce that some seedlings need watering because similar plants previously withered in a comparable situation. In cases where no prior experience exists—such as with pest attacks on unfamiliar crops or new cultivation methods—participants may need to take risks and test an "operation" to see if it produces satisfactory results. If uncertainty remains, the director can conduct a trial and share the results with the participants. In these instances, participants are expected "to confirm the results of their director's trial" (Concept-27).

At a more advanced stage, participants must decide what to observe independently. If unsure, it may be helpful "to deduce general principles from previous instructions on various 'operations'" (Concept-28). For example, a participant may learn that tubers and bulbs require less frequent watering than leafy vegetables, a guideline that can inform future decision-making, even if it is initially imprecise.

Subcategory "Utilization of Opportunities to See Things Ordinarily Invisible"

Providing participants with opportunities to observe what is typically hidden can enhance their ability to infer overall crop conditions from visible features. A primary concept in this subcategory is "to examine the underground situation" (Concept-29). Participants gain valuable insights by examining the root systems of crops and weeds, which they can observe when turning over ridges after harvesting. This practice helps them estimate the extent of root growth, improving their ability to water crops effectively. For example, one participant initially watered only the base of each crop, assuming the roots extended only a short distance, based on their previous experience of seeing rice stubbles with roots shortly cut off in paddy fields. However, after observing actual crop roots, this misconception was corrected.

Another concept in this subcategory is "to examine the interior and taste of crops" (Concept-30). This is exemplified in the harvesting of the *Pisum sativum* variety 'usui endo,' whose unripe green peas are typically boiled with rice. Participants learned to discern which pods were in optimal condition for harvesting by correlating pod appearance with the taste of the peas inside. Wrinkled pods indicate that the peas are overripe, bitter, and less sweet. As participants learn to understand the relationship between external appearance and internal quality, they can better evaluate the quality of crops based on visual inspection.

Subcategory "Utilization of Unconcerned Opportunities"

Participants have opportunities to observe every aspect of crop growth in their daily lives. They can imagine the full shape of a crop plant if they have engaged in the entire farm tasks to raise the crop. Not only that, however, they can also observe the whole appearance of some leafy vegetables on the store shelves. By employing "unconscious observation" (Concept-31), participants can maintain an appropriate distance between roots and ridges, allowing adequate space for crops to expand during seed sowing, seedling transplanting, staking, and providing coverage for young plants.

When participants follow instructions, they may assume that the director has a more precise understanding of the tasks at hand. However, it is often more beneficial for them to make "observations" during "operations," as they are closer to the task and its details. As such, "confidence in being more informed than the director" (Concept-32) encourages participants to critically examine the instructions, which can lead to more appropriate decisions based on their current circumstances.

Rather than the director instructing participants to observe each work object, a time-saving strategy is "habitual observation of the entire farm" (Concept-33). By doing so, participants recall past actions, such as "when this happened, we took that action" or "after we completed this task, that change occurred." Even in the absence of specific guidance, they may recognize the need to act when unexpected issues arise with crops or the environment.

3.3.2.6 Components of the Category "Decision-Making" (Four Concepts)

These concepts are prerequisites for understanding what "operation" follows next without waiting for the director's instructions. When participants form a cyclical process involving "decision-making," "observation," and subsequent "decision-making" based on observations, they gain autonomy from the director.

The primary goal of "observation" is "to determine whether a particular 'operation' should be performed at a specific time" (Concept-34). For example, when participants ask the director whether they should water a particular crop, the director typically provides observational results. However, moving forward, the director may use a subjunctive expression such as, "You should water the crop if there is a risk of it wilting or facing other issues," prompting participants to observe the crop's current condition and make the decision themselves.

Many participants initially challenged this concept. An instructor from an employment assistance program, mentioned earlier, argued that the goal of job training was merely to teach trainees to follow instructions. Similarly, other participants believed they were simply expected to follow the director's guidance. In a conversation during a morning meeting, the following exchange occurred among some regular participants, including me:

Author: You should observe the crops to understand their needs.

Participant A: They tell us nothing.

Participant B: When we decide what to do, you get angry and say, "Don't work in a self-serving way."

Participant C: I understand what you mean—you want us to think for ourselves, right?

Author: No, I mean you should observe the crops, not just think on your own.

Participant A: You should observe the crops and tell us what they need.

The participants' statements reveal the director's role in conveying instructions from the crops themselves. To become independent from the director, participants need "to stop expecting instructions and observe the crops instead" (Concept-35). At times, I needed to clarify what actions participants should take, so they could carry out "operations" based on their observations. This often involved walking around the farm with the participants, highlighting the crops' current status, and guiding them to prioritize a limited set of feasible actions. In other words, I encouraged participants "to prioritize a limited number of 'operations' based on 'observation' results" (Concept-36).

Of course, participants may sometimes misjudge priorities. I hoped that they would seek feedback on their decisions, whether right or wrong. However, one participant expressed a reluctance to take responsibility, stating, "I will be blamed if my decision is wrong, so I make it a rule to follow the director's instructions." On another occasion, when discussing the timing of opening and closing vinyl covers—due to conflicting goals of preventing frost damage and bolting—another participant showed indifference, saying, "I don't want to deal with such a complicated issue." This illustrates the challenge of encouraging participants to make observations when they fear failure. It is essential to emphasize the need "to learn from failures" (Concept-37).

3.4 Discussion

3.4.1 The Present Attainments

The series of "changes in things to be learned during the increase in autonomy of skill acquisition" represents a two-phase transformation of the tripartite relationship between participants, their director, and crops (Fig. 3.4). Initially, participants follow the director's instructions, requiring the director to first observe the crops and environment (a). After a period, participants gain independence from the director's guidance (b). This marks the first phase of transformation, allowing participants to anticipate instructions based on their past experiences. However, the relationship between participants and crops remains unilateral at this stage. In the second phase, participants learn to read signs from their crops or "take directions from crops" (c). This renewal of the relationship allows participants to make decisions based on their own observations of the crops, and by extension, the environment. Notably, while many participants successfully navigate the first phase, they often hesitate to engage in the second, which is crucial for initiating autonomous farming activities.

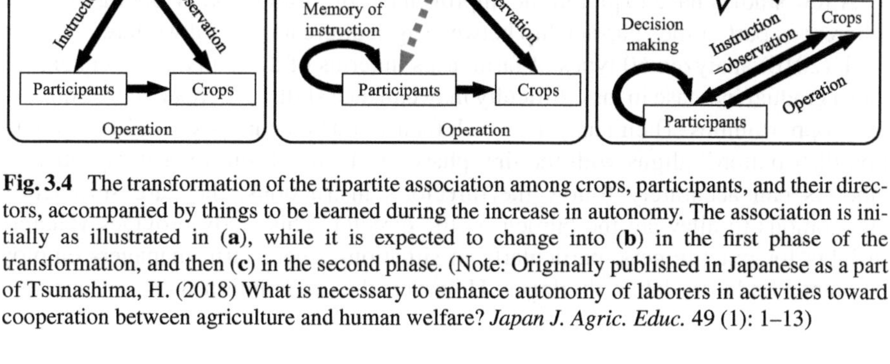

Fig. 3.4 The transformation of the tripartite association among crops, participants, and their directors, accompanied by things to be learned during the increase in autonomy. The association is initially as illustrated in (**a**), while it is expected to change into (**b**) in the first phase of the transformation, and then (**c**) in the second phase. (Note: Originally published in Japanese as a part of Tsunashima, H. (2018) What is necessary to enhance autonomy of laborers in activities toward cooperation between agriculture and human welfare? *Japan J. Agric. Educ.* 49 (1): 1–13)

The first phase of change mirrors efforts to enhance adaptation to the existing labor market. Consider a participant as a trainee using employment support services. Initially, they are assigned tasks according to the director's instructions, consistent with the goals set by employment service providers. Although these tasks may seem rudimentary, they can be highly challenging, particularly for trainees with developmental disabilities. Their success at this stage is regarded as merely preparatory for actual job hunting, as noted in interviews with job training instructors in the previous chapter. True employment activities begin later, where the role of employment support organizations begins to resemble career education, commonly practiced in high schools and colleges. At this point, instructors increasingly criticize employees who merely wait for instructions, a mindset that stems from the belief that employees should follow directions. Advocates of career education, such as Yanagida (2015), stress the importance of future employees taking initiative in their workplaces, consistent with the concept of "basic social skills" (Ministry of Economy, Trade and Industry 2017). This concept emphasizes taking ownership of tasks, rather than passively awaiting instructions, aligning with "human ability" promoted by the Ministry of Education, Culture, Sports, Science and Technology (see Chap. 2).

At the present farm, some participants, particularly elderly homeless laborers, demonstrated a proactive attitude, anticipating instructions rather than waiting for them. These individuals often shared experiences, such as their time on construction sites, where no explicit guidance was provided. Instead, they learned by observing experienced coworkers and applying trial-and-error methods. In contrast, younger participants seemed to wait for instructions. These older laborers had internalized the concept of "basic social skills" enough to take initiative, yet they still resisted "taking directions from crops," focusing instead on inferring the director's intentions. The notion of "basic social skills" assumes that directions come from individuals or organizations with substantial influence over one's work, such as superiors or customers. In contrast, "taking directions from crops" involves reading signs from plants, over which growers exert complete control. Thus, the second phase of change requires a shift toward complete independence from powerful individuals while fostering dependence on seemingly powerless entities—an essential transition from loyalty to the former to attentiveness to the latter. The difficulty in entering this phase may stem from the prevailing societal belief that directions must flow from the powerful to the powerless.

A few studies have explored the improvement of farming skills among disabled laborers as part of the "partnership between agriculture and social welfare." Toyoda et al. (2016) analyzed 60 types of farm tasks in terms of difficulty. They concluded that a gradual increase in task difficulty may enhance skills and create more employment opportunities. Their analysis, which focuses on "manual dexterity" and "attention distribution," aligns with the first phase of change. From their perspective, if laborers can accurately follow the director's instructions, they are sufficiently autonomous to alleviate the burden on the director. This study provides valuable insights for practitioners, particularly in situations where farm managers employ laborers to address labor shortages. However, this assumption does not apply to cases where city dwellers attempt to utilize abandoned farmland without prior farming experience or employment with professional farmers. In these cases, the partnership takes a different form, one that does not revolve around providing disadvantaged laborers with opportunities in existing farms. While the second phase of change may be more demanding than the first, as discussed earlier, it remains a practical and valuable pursuit.

3.4.2 Past Discussion on What It Is to "Take Directions from Crops"

As noted earlier, the current action research on agriculture-related job creation unexpectedly encountered the challenge of "taking directions from crops." This concept refers to a gap in the lessons learned from previous research. To clarify the problems addressed in the subsequent chapters, this section examines existing discussions on the relationship between people and cultivated plants.

Firstly, Tsuno (1995) described "taking directions from crops" in the context of traditional agriculture as follows:

> Farmwork involves a cyclical process of interpreting what crops demand, which growers then address through their labor. Afterward, they assess whether they have met the crops' needs. Rather than passively observing crops, growers compare those that have received particular treatments with those that have not. Through this process, crops emit various signals—such as the gloss of leaves, the posture of stems, and the way blossoms bloom—that appeal to the human aesthetic sense.

According to Tsuno, the "aesthetic sense" of growers is the receiving apparatus of "directions" from crops. While he does not fully explain how this sense functions, he suggests that agriculture, shaped by "labor driven by fear of death," led to an understanding of nature's laws and developed agriculture as a means of harnessing nature's power. This implies that growers develop the "aesthetic sense" through their fear-induced labor.

Secondly, Matsuo (2005) also emphasized the cyclical nature of horticulture, characterized by "observation," "decision-making," and "operation." He argued that "growers keenly detect what plants demand and fulfill these needs, while plants respond to the treatments they receive and silently indicate what they require next." Matsuo contended that it is impossible to teach the acquisition of the "aesthetic sense." He argued that traditionally, only compelling forces motivate people to engage in horticulture, leading them to learn the necessary skills through repeated cycles of interaction. Moreover, these "compelling forces," which may be exemplified by the above "fear of death," are essential for cultivating the "aesthetic sense." However, such forces could be detrimental to potential horticulturists' well-being, contradicting the goals of social welfare.

Thirdly, Toyohara et al. (2007) advocated for the use of horticultural activities to enhance participants' well-being. They explained that "people are 'fascinated' by the plants they grow when they observe the plants' responses to their care, which motivates continued engagement in horticulture." This view assumes that people already possess the "aesthetic sense," which allows them to recognize the plants' responses. If this assumption holds, horticultural activities may uncover the latent abilities of vulnerable individuals and lead to their empowerment. However, as previously discussed, this assumption is not universally applicable; some individuals may not be "fascinated" by the plants they tend, which limits the effectiveness of horticultural activities.

Following these discussions, Tsuno (2005) argued that agriculture's appeal lies in the "touching symbiosis between humans and crops or livestock." This "symbiosis" mirrors the relationship between growers, who provide care, and plants, which in turn fascinate the growers. Tsuno posited that the heartwarming properties of these symbiotic relationships, rather than "compelling forces," effectively foster the "aesthetic sense." Once growers acquire sufficient knowledge and skills to recognize their symbiotic relationships with crops, positive feedback can promote further learning, establishing a virtuous cycle. From the perspective of educational engineering, this approach appears valid. However, Tsuno did not address the possibility that some individuals may remain untouched by such symbiosis. He overlooked the question of what happens when potential horticulturists are unable to acquire the "aesthetic sense" due to an inability to connect with crops. Practically speaking, an essential prerequisite for forming this connection is a certain sensitivity—the "aesthetic sense." This raises a circular argument: a form of "aesthetic sense" is necessary to develop the "aesthetic sense."

Of the four references by Japanese authors discussed above, the first two suggest changes in the attitudes of autonomous growers toward nonhuman life forms, while the latter two assume that the "aesthetic sense" is innate in all participants. In this regard, the first two align with the findings from the previous subsection, while the latter two do not. Meanwhile, Western research on green care aligns more closely with the latter two perspectives, with studies indicating that farming activities enhance clients' well-being once they have successfully followed "directions" from crops or other nonhuman beings. For instance, Kogstad et al. (2014) observed that it was important for clients to connect with beings, such as farm animals, who depend on their care and reciprocate without favoritism. Regarding how to "take directions from crops," they noted that "the ability to connect with nature may differ depending on childhood experiences" and emphasized the importance of allowing children to build this connection. Garcia-Llorente et al. (2018) shared a similar view, asserting that interactions with nature promote well-being for people at risk of social exclusion. However, these articles do not explore how to cultivate such an ability, and some even suggest it may be impossible for adults to acquire it.

This discussion, informed by the previous section, highlights two key points: First, many participants struggle to acquire or demonstrate the "aesthetic sense," which is essential for autonomous activities. Second, there is no scientifically developed method to cultivate the "aesthetic sense" without relying on "compelling forces," in other words, with the spontaneity of prospective growers. Moreover, even "spontaneity" can hinder the development of the "aesthetic sense" unless it evolves from a focus on power to an appreciation for powerless life forms. Consequently, the question of how to "take directions from crops" remains largely unaddressed in prior research. A more thorough examination will be undertaken in Chap. 5, where a precise definition of "aesthetic sense" will be provided. Meanwhile, we can still seek for an alternative to the "compelling forces" in this context.

3.4.3 Results of Additional Interviews

As discussed in the previous section, many participants struggled with the concept of "taking directions from crops." They appeared to assume that there was an instruction manual they could rely on, but only I, as the director, had access to it. In reality, no such manual exists. My role was, therefore, to "take directions from crops" through my own observations and decisions, acting on their behalf and hoping they would eventually develop the ability to do so independently. This subsection provides background on this issue, particularly in relation to their attitudes toward work.

As noted in the final paragraph of the second section, I reviewed the results of additional interviews, from which I extracted relevant insights. Interviewees openly discussed their views on wage labor, shaped by their complex experiences as day laborers, including frequent job changes prior to entering the day labor market.

The first point to address is why the new participants applied for the job offered in this action research project. None did so enthusiastically; one interviewee even stated that he disliked farmwork. A core belief among the homeless laborers was a refusal to receive welfare benefits. However, they faced challenges in finding alternative employment and were compelled to earn a living on their own. This necessity led them to seek job opportunities and to learn through experience, even if the potential for diversified income was limited. Farmwork, thus, became a viable option.

Several interviewees expressed feeling restricted in their job choices. One relatively young day laborer in his 40 s sought opportunities outside the construction industry, where he had grown weary of his work. He explained that supervisors often abused him verbally, making demeaning remarks such as, "Can't you perform such an easy task like this?" He also noted that day laborer brokers in Kamagasaki exploited the vulnerability of homeless workers, taking advantage of their uncertain employment situation.

Another day laborer in his 60 s complained of the physical limitations that came with age, making it difficult to work as he once did. He was frustrated by being excluded from the labor market due to his age, even though he still wished to remain active. Despite this, he expressed patience with both the physical demands of his labor and the humiliation of being despised and exploited. Similar sentiments were shared by other homeless interviewees in their 60 s, who commonly mentioned "patience." They emphasized their willingness to accept any work assigned to them, convinced that following instructions and receiving wages would eventually lead to mastery of the job. As a result, these senior interviewees accepted any opportunity that came their way, maintaining that they would avoid unemployment only if they did not remain overly particular about the work.

Regarding how they learned their duties, interviewees shared various guiding principles based on their experiences, such as "The trick of learning is taking to your present job," "By trial and error," "You must learn by observing your colleagues or seniors," "You must imagine what the results will be," "You must devote yourself to your present job," and "You must devise your own way of learning, such as taking notes." These interviewees expressed a desire for opportunities to acquire skills that could lead to earned income. Some shared their aspirations as follows:

> It would be interesting for me to learn about the entire operation of the business so that I can follow it long-term.

> I will master the entire operation of any business once you teach me how to conduct it. I want someone to teach me.

They also specified that their instructors should provide detailed guidance, noting that it would be "torture" to be left "imperfectly informed" or "completely at a loss" during on-the-job training. Additionally, they even requested that coaches inform them in advance of the tasks they would be expected to perform. In summary, there was a tacit understanding among participants that they needed someone to guide them meticulously, leaving no room for mistakes or contradictions with supervisors. This mindset reflects their continued reliance on authority figures, maintaining the traditional superior-subordinate relationship. They submitted themselves to this dynamic, which was still unperturbed by nonhuman elements contrary to the previous chapter.

Regarding the working environment, several interviewees expressed a preference for positive human relationships, stating that "It would be best if colleagues can get along well with each other," "It is pleasant to work in a group," and "I hate any quarrels at work," all of which align with their dislike of humiliation. One interviewee jokingly mentioned that "selling the vegetables we have harvested is fun because we can meet local *madams*," suggesting a desire to form ties outside of their immediate workplace. This reflects an anthropocentric outlook, wherein they sought to serve the local community (Fig. 3.5). A retired day laborer in his 70 s, who received welfare, repeatedly remarked, "Since I am making my livelihood with support from taxpayers, I have to contribute to society in return."

3.4.4 Implications of the Additional Interviews and Supplements

The previous subsection addressed the interviewees' ascetic attitude toward paid work, which is rooted in their perception that wage labor involves following the instructions of more powerful labor market actors, such as supervisors. From their perspective, it may be more rational to follow these instructions rather than "take

Fig. 3.5 Selling vegetables at a public space in a residential neighborhood

directions from crops," as doing so could risk their peace of mind by opposing their superiors. They often anticipated instructions even by reading their director's expressions at the farm. Although autonomous activities might improve well-being, as noted earlier in this chapter, some interviewees believed it outrageous to seek pleasure through wage labor, let alone "fascination," given that they are paid for toil. This sentiment was particularly evident among homeless elderly individuals who had previously worked as regular employees or day laborers. It appears they were eager to restore the commercial value of their labor, which had been denied for years, regardless of the exploitative and unpleasant nature of the employment. Consequently, homeless elderly laborers rarely seemed to feel liberated during farmwork, unlike the younger trainees receiving employment assistance discussed in the previous chapter. This observation also clarifies why the interviewees, particularly the senior homeless laborers, did not consider nonhuman beings such as crops worthy of "giving directions." The following subsection will report on a subsequent attempt to address this issue.

Some younger interviewees revealed distinct attitudes toward farming that somewhat diverged from the aforementioned difficulties in "taking directions from crops." While these examples may not significantly affect the chapter's overall conclusions, they serve to demonstrate that the interviewees were not a monolithic group and that their views on farmwork could evolve with experience. These cases will be referenced in later chapters.

First, a welfare recipient in his 30s joined the farming activity based on the advice of his employment assistance caseworker. A few months after sowing

Fig. 3.6 Radish sprouts sowed by the interviewee to be used for food-serving in Kamagasaki

radishes, he found himself excited by their growth and harvest. He discovered that, unlike the industrial production in which he had previously worked, agriculture required time for products to materialize. He realized that he could enjoy the slow pace of agricultural work, which seemed to suit him. He noted, "Before becoming involved in the farming activity, the purpose of any job I had was purely financial, but now my perspective is changing."

Second, another welfare recipient in his 30s, initially indifferent to agriculture, began searching for agricultural jobs after being encouraged by his caseworker. Reconnecting with his rural background, he recognized that outdoor work suited him better than working in cramped urban conditions. He found it encouraging to see customers purchasing the vegetables he had grown with his colleagues.

Third, a homeless laborer in his 40s, like other homeless participants, actively applied for the current job opportunities. He emphasized that his enjoyment of learning new things made the work not solely financial. Growing and harvesting crops led to a complete shift in his perspective (Fig. 3.6). He hoped to continue this work as a hobby, since he was unsure about the profitability of his produce. Senior homeless interviewees also mentioned the unexpected joy of harvests and good sales. However, one participant expressed dissatisfaction with the tasks of picking and packaging vegetables, which required precise finger work, and found conversing with customers burdensome.

3.4.5 Additional Attempt at "Taking Directions" from Consumers

Returning to the earlier discussion in the previous subsection, the interview results suggest that there remains a possibility for participants to "take directions" from other humans—specifically, consumers. This warrants consideration of the act of selling farm produce. As mentioned in the previous chapter, we sold our products at various venues around Osaka City to ensure the success of the action research project as a business-oriented job creation initiative. Although not included in the M-GTA analysis as an off-farm activity post-harvest, sales inevitably involved communication between shoppers and participants. When participants acted as salespeople in face-to-face encounters, they often received questions about product availability and feedback on previous purchases.

The most common consumer feedback was: "The product I bought before was delicious; I want the same one again." Among various potato varieties, one selection, the 'red moon,' gained attention for its reddish skin looking similar to sweet potatoes and favorable taste, resulting in repeat customers. Similarly, consumer feedback helped guide the selection of other crops, such as pumpkin and sweet potato. Additionally, when we failed to produce head-forming Chinese cabbage or it bolted prematurely, we decided to sell the sprouts with flower buds, which were well received for their sweeter taste compared to ordinary rape flowers. In this case, consumer input also guided our decision on what to sell.

Some questions from shoppers left participants embarrassed. The most common inquiries involved cooking methods for certain products. For example, when asked about the 'red moon' potato variety, participants struggled to respond due to their limited knowledge of cooking. To assist them, I later posted a point-of-purchase advertisement stating, "This potato variety is difficult to boil." Afterward, a customer inquired whether the variety was more akin to Irish Cobbler or May Queen, two potato variety groups known for their distinct boiling characteristics. The participant, unaware of this distinction, attempted to divert the conversation to the cultivation method. After I clarified the customer's question, the participant remarked, "I didn't know it's so difficult to sell vegetables," though he later offered fried potato samples with an explanation of the variety (Fig. 3.7).

Participants who had spent long periods in shelters for the homeless or day laborer flophouses had not previously had the opportunity to cook for themselves and learn how consumers typically categorize potato varieties. Additionally, crops that were easier to grow, such as marrow squash (*Cucubita pepo* 'kinshi-uri') and winged bean (*Psophocarpus tetragonolobus*), were unfamiliar to consumers. Curious shoppers often asked participants about how to cook these items. When participants could not provide adequate answers, potential customers were discouraged from purchasing. These incidents prompted participants to taste their own produce in their spare time to better understand how consumers might prepare the items. This knowledge was later applied in subsequent sales opportunities.

In summary, communication with consumers reveals their needs for both goods and information, encouraging participants to adjust their crop selection and present their products according to customer preferences. The findings further suggest that consumers play a role in the tripartite relationship between crops, participants, and their director. As consumers cook, taste, and digest the farm produce, their opinions are shaped by natural processes, not human arbitrariness. Continued consumer involvement may enable them to assume a role similar to that of the director, thus allowing participants to gain greater independence from their current supervisor (Fig. 3.8).

a

b

Fig. 3.7 Selling and promoting a potato variety 'red moon.' (**a**) Fried samples presented next to the packages for sale. The POP says, "hard to boil down and very sweet." (**b**) Two homeless participants offer the samples to passers-by

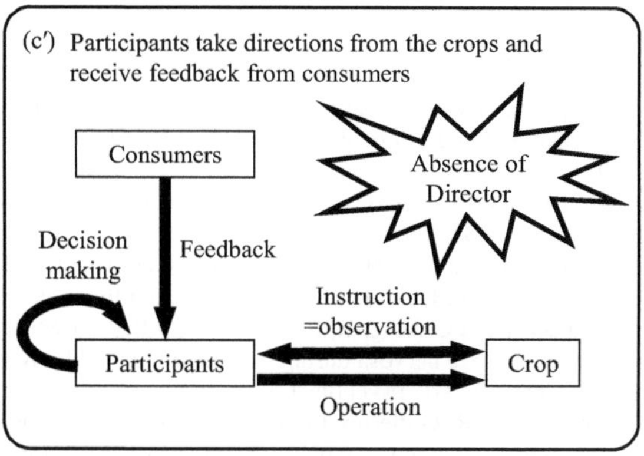

Fig. 3.8 Another possibility of the tripartite association among crops, participants and their director to replace the diagram (c) in Fig. 3.4

3.5 Conclusion

The answer to the question posed in the introduction can be summarized as follows: The relationship between the instruction giver and receiver in the current context is not definite; rather, it should be restructured, particularly to include both human participants and nonhuman life forms, from the perspective of autonomy. The analysis, through M-GTA, introduced the phrase "take directions from crops" to challenge traditional thinking about power dynamics. This personified expression proved valuable in describing the intrinsic nature of agricultural labor and revealed that participants, especially elderly homeless laborers, often subordinated themselves to their director.

However, by the time data collection was completed, the theory outlined in this chapter proved impractical from the participants' viewpoint. They consistently avoided "taking directions from crops," an action central to the theory of farming. As a result, participants missed opportunities to engage meaningfully with crops, treating farmwork instead as wage labor or part of employment training. Initially, this initiative sought to create alternative job opportunities outside the conventional labor market. Consequently, I found the instruction giver-receiver relationship on the farm somewhat awkward. It persisted longer than I expected, as participants, despite their status, struggled to detach from the notion that they were subject to their director's authority. My attempts to dissolve this hierarchical relationship with the human director were unsuccessful. Nevertheless, there remains hope that ongoing relationships with consumers may help overcome this challenge.

In conclusion, it seems contradictory to pay participants wages while simultaneously offering opportunities for enjoyment in farmwork. In retrospect, one possible reason for this contradiction is that I failed to effectively communicate to participants the essence of growing crops. The next chapter will explore an alternative approach to addressing this issue.

References[3]

Bellows AC, Alcaraz GV, Vivar T (2010) Gardening as tool to foster health and cultural identity in the context of international migration: attitudes and constraints in a female population. Acta Hortic 881:785–792

Cabannes Y, Raposo I (2013) Peri-urban agriculture, social inclusion of migrant population and Right to the City. City 17(2):235–250

Elings M (2006) People-plant interaction: the physiological, psycological, and sociological effects of plants and people. In: Hassink J, van Dijik M (eds) Farming for health. Springer, pp 43–55

Fendel U, Sandler H, Papachristou C, Voigt B, Rose M, Klapp BF (2018) Bodily experiences of patients diagnosed with anorexia nervosa during treatment with the body monochord: a modified grounded theory approach. Arts Psychother 59:7–16

Garcia-Llorente M, Rubio-Oliver R, Gutierrez-Briceno I (2018) Farming for life quality and sustainability: a literature review of green care research trends in Europe. Int J Environ Res Public Health 15(6):1282

Gasperi D (2015) How can UAG contribute to make the residential areas from 1960s and 1970s more liveable? In: COST Action TU 1201 "Urban Allotment Gardens in European Cities" short report of the short term scientific mission

Haller RL (2017) Goals and treatment planning: the process. In: Haller RL, Capra CL (eds) Horticultural therapy methods: connecting people and plants in health care, human service, and thearpeutic programs, 2nd edn. Taylor and Francis, New York, pp 27–36

*Iiba S, Yamabata N (2022) Actual situation and problems of agricultural employment support activities for persons with disabilities. J Rural Probl 58(3):157–164

Jackson B, Marshall M, Schofield S (2017) Barriers and facilitators to integration of physician associates into the general practice workforce: a grounded theory approach. Br J Gen Pract 2017:e785–e791

Kambaru A (2018) Qualitative research and a modified grounded theory approach. Tsuru Univ Rev 88:47–58

***Katakura K, Yamashita H, Kiyomitsu K (2007) Employment of the disabled in farm management [Nōgyōkēē ni okeru shōgaishakoyō no manejimento]. J Rural Probl 43(1):78–83

***Kinoshita Y (1999) Grounded theory approach: restoration of qualitative empirical research [Guraundeddo-seorī-apuriōchi: Sitsutekijisshōkenkyū no Saisē]. Koubundou, Tokyo

***Kinoshita Y (2003) Practice of grounded theory approach: introduction to qualitative research [Guraundeddo-seorī-apuriōchi no Jissen: Sitsutekikenkyū he no Izanai]. Koubundou, Tokyo

***Kinoshita Y (2005) Grounded theory approach: practical edition for different fields [Bunyabetsu Jissen-hen Guraundeddo-seorī-apuriōchi]. Koubundou, Tokyo

***Kinoshita Y (2007) A live lecture: all about practical qualitative modified grounded theory approach [Raibukōgi M-GTA: Jissentekishitsutekikenkyūhōshūsēban Guraundeddo-seorī-apuriōchi no Subete]. Koubundou, Tokyo

Kogstad RE, Agdal R, Hopfenbeck MS (2014) Narratives of natural recovery: youth experience of social inclusion through green care. Int J Environ Res Public Health 11:6052–6068

*Kotani K, Naito Y, Shimada M, Kojima H, Sasada K (2016) Current situations and issues of the employment of disabled people by agricultural corporation. J Jpn Soc People Plant Relationships 16(1):23–28

***Matsuo Ē (2005) An exhortation to social horticultural science [Shakaiengēgaku no Susume]. Rural Culture Association Japan, Tokyo

McAdam R, Mason B, McCrory J (2007) Exploring the dichotomies within the tacit knowledge literature: towards a process of tacit knowing in organizations. J Knowl Manag 11(2):43–59

[3]*: in Japanese with an English title and summary, **: in Japanese with an English title, ***: in Japanese.

***Ministry of Economy, Trade and Industry (2017) Basic social skills, 3 Feb 2017. http://www.meti.go.jp/policy/kisoryoku/. Accessed 11 Dec 2017

Moruzzo R, Di Iacovo F, Funghi A, Scarpellini P, Diaz SE, Riccioli F (2019) Social farming: an inclusive environment conducive to participant personal growth. Soc Sci 8:301. https://doi.org/10.3390/socsci8110301

Nagayama Y, Hasegawa M (2014) Nursing care process for releasing psychiatric inpatients from long-term seclusion in Japan: modified grounded theory approach. Nurs Health Sci 16:284–290

**Nakamoto E, Sawano K (2020) The effects of universal agriculture and JGAP on expanding work fields of persons with disabilities. Jpn J Farm Manag 58(3):21–26

Okura M (2019) The process of structuring community health needs by public health nurses through daily practice: a Modified Grounded Theory Study. Asian Nurs Res 13:229–235

Opitz I, Berges R, Piorr A, Krikser T (2016) Contributing to food security in urban areas: differences between urban agriculture and peri-urban agriculture in the Global North. Agric Hum Values 33:341–358

Shen W, Yuan Y, Liu C, Luo J (2016) In search of the 'Aha!' experience: elucidating the emotionality of insight problem-solving. Br J Psychol 107(2):281–298

Tei F, Gianquinto G (2010) Origin, spread and multifunctional role of the urban horticulture. Italus Hortus 17(1):59–73 (in Italy with English summary)

Torquatti B, Stefanib G, Massinic G, Cecchinia L, Chiorria M, Paffarinia C (2019) Social farming and work inclusion initiatives for adults with autism spectrum disorders: a pilot study. Wageningen J Life Sci 88:10–20

*Toyoda M, Kaneko M, Yokota Y, Asai S, Fudano T, Shiroyama Y (2016) A new assessment method of agricultural work in job assistance for people with intellectual disabilities. J Jpn Soc People Plant Relationships 15(2):1–10

***Toyohara N, Yōichi I, Kae M (2007) Agricultural and horticultural activities for well-being: a manual of how to practice effortlessly [Fukushi no tame no Nōengēkatsudō: Murisezu Dekiru Jissen Manyuaru]. Rural Culture Association Japan, Tokyo

*Tsunashima H (2018) What is necessary to enhance autonomy of laborers in activities toward cooperation between agriculture and human welfare? Jpn J Agric Educ 49(1):1–13

***Tsuno Y (1995) Small agriculture: a quest from mountainous farm villages [Chīsai Nōgyō: Sankanchinōson kara no Tankyu]. Rural Culture Association Japan, Tokyo

**Tsuno Y (2005) Reconsideration of agro-technology through the development of weed control in paddy field by mulching using the recycled paper and the non-woven fabric, made by scraped cotton. J Weed Sci Technol 50(4):300–309

*Yanagida J (2015) Case study of university students' peer group activities in the career development processes: from developmental network and mental abilities in socialization perspectives. Tokyo Univ Inf Sci Fac Inform 19(1):59–86

Chapter 4
Learning from Other Cases in Japan: Further Challenges in the Partnership Between Agriculture and Social Welfare Sectors

Abstract To examine alternative interpretations of the "instruction giver-receiver relationship," I analyze eight cases of partnerships between agriculture and social welfare, in addition to two cases of personal autonomous gardening. For the eight partnership cases, I conducted semi-structured interviews with the directors of farming activities, who were staff members of support organizations responsible for instructing clients or trainees. For the two independent gardeners, I conducted unstructured interviews to explore their process of skill acquisition. The interviews revealed that the directors did not explicitly expect their participants to be autonomous. Instead, they stated that clients should only do what they were capable of. Some cases involved participants who were independent of directors either constantly or occasionally, but the directors, as well as the participants themselves, could not articulate the nature of this autonomy. This lack of clarity also applied to those independent gardeners. The key lesson drawn from this chapter aligns with the well-known proverb: "What one likes, one will do well." By considering these findings from a different perspective, the only individual who emphasized the significance of autonomy was myself. The rationale behind this emphasis is simple: no one should be deprived of the opportunity to "take directions from crops."

Keywords Autonomous farming/gardening · Employment support for disabled people · Instruction giver-receiver relationship · Skill acquisition · Vicariousness

4.1 Introduction

4.1.1 Search for an Alternative "Instruction Giver-Receiver Relationship"

The previous chapter analyzed the "instruction giver-receiver relationship" observed during my action research project aimed at job creation on a farm. It concluded that communication with participants could have been more effective. The goal of the current chapter is to explore alternative models for this relationship. To address this,

H. Tsunashima, *Farming and Social Inclusion in Japan*, Urban Agriculture,
https://doi.org/10.1007/978-3-032-01598-3_4

I adopt a dual approach: first, by examining cases of "partnerships between the agriculture and welfare sectors," and second, by considering situations where no human instructor is involved. As discussed in the previous chapter, past research on such partnerships assumes that relationship formation natural, while studies on urban horticulture in the West report instances of autonomous skill acquisition, which supports self-sustained farming activities. In April 2017, after completing the M-GTA-based analysis in the previous chapter, I encountered an elderly day laborer from Kamagasaki who, while working, also enjoyed gardening in a suburb of Osaka. I believed his gardening could exemplify autonomous skill acquisition and, more conveniently, could be studied without international travel. Unlike partnership-based cases, this example is linked to Japan's long history of using agricultural resources for poverty alleviation, a topic explored in the next subsection.

4.1.2 History of Natural or Agricultural Resource Usage for Poverty Alleviation

During the Nara period (710–794), the imperial court planted persimmon trees along post-station routes. A report to the throne by a Buddhist priest affiliated with Todaiji Temple, who had served as an ambassador to the Tang court. Inspired by the then capital, Changan, the report suggested these trees could provide shade and food for travelers in need. While the court continued planting fruit trees during the Heian period (Kodama 1986), during the Edo period (1603–1867), the government favored conifers and deciduous trees, which were easier to maintain (Takebe 1988). Today, while fruit trees are rarely planted along roadsides in Japan, they are considered a potential resource for urban poverty alleviation globally (cf. Kowalski and Conway 2018; Kazemi et al. 2018).

In a village in Okinawa Prefecture, needy people historically collected wild shekwasha (*Citrus depressa*) from common areas, a practice that persisted until the commercial boom of Okinawan products altered the fruit's role (Sugihara 2008, 2010). On Miyakojima Island, residents maintain kitchen gardens near their homes, growing fruit trees, vegetables, herbs, and flowers for personal use and community sharing. However, the Ministry of Agriculture, Forestry, and Fisheries overlooks the significance of these kitchen gardens in assessing Okinawa's food self-sufficiency rate, a view inconsistent with local perspectives (Sugihara and Shinzato 2010).

Regarding domesticated animals, a historical example involves Okinawan immigrants who could not afford to emigrate abroad after mass unemployment struck Okinawa in the 1920s. Many settled in coastal industrial areas in Kansai and Kanto. In Amagasaki City, Hyogo Prefecture, they built huts in a riverside forest along the Muko River, where they had been engaged in levee construction, eventually raising poultry. Despite a municipal eviction order, the area became home to around 300 people (Ishihara 1982). The scene described is as follows:

> On the thatched roofs, the green leaves and bright flowers of pumpkins, sponge gourds, and moonflowers are in full bloom, attracting butterflies and bees. Such a mild view is rarely seen in city centers… Most male laborers work temporarily in Kobe or Osaka and raise pigs. At their peak, each household rears 40–50 pigs, requiring women and children to gather food scraps from nearby cities like Takarazuka and Nishinomiya… Even during the post-WWII famine, the community rarely went without food, and some even accumulated wealth (Ishihara 1982).

In contemporary Western terms, Hardman and Larkham (2014) would categorize this instance as "guerrilla gardening," a term they use to describe the unauthorized cultivation of vacant urban plots. The legal ambiguity surrounding such activities may have attracted academic attention. Furthermore, discussions on the significance of urban gardening, particularly those focusing on environmental concerns, have gained increasing prominence. However, recent studies often overlook the fact that the contemporary rise of urban gardening since the 1980s was a response to the "new poverty" and "social exclusion"—an issue notably underlined in France. This movement represents a renewal of the nineteenth-century "workers' garden movement" seen in some industrialized European countries (cf. Nilsen and Barnes 2014; Muramatsu 2018).

Urban gardening aimed at poverty alleviation is closely tied to the enhancement of gardeners' skills, a consideration particularly relevant for informal trades. Additionally, immigrant laborers from Okinawa, who had been managing kitchen gardens like those seen in Miyakojima, with additional pig farming, exemplify the transfer of skills from their home regions. This is comparable to the urban farming cases of immigrants described by James (2016) and the studies cited in the previous chapter (e.g., Gianquinto 2010; Bellows et al. 2010). It is evident that these immigrant farmers had achieved a degree of skill autonomy in their agricultural practices.

Should these horticulturists still reside in their current environment, sourcing daily necessities from the surrounding natural landscape rather than relying on formal employment, they would likely alleviate concerns about unemployment or job instability. In doing so, they could gain greater confidence in the labor market. Furthermore, if similar investigable cases still exist today, they could serve as models of autonomy for participants in job creation programs, as seen with the aforementioned day laborer. However, my search across Japan yielded no comparable example, aside from those mentioned below.

4.2 Methods of Case Investigations

For cases within the agriculture-welfare partnership category, I identified potential cases through word-of-mouth communication with visitors to the aforementioned project farm and members of practitioners' communities. The selection criteria were designed to ensure the study reflected the diversity of enterprises practicing the partnership. First, since most practitioners exclusively employed laborers with disability certificates, I aimed to include enterprises that employed individuals without these certificates. Second, to explore the factors that contribute to the autonomy of laborers in farming activities, I sought to include cases where laborer autonomy varied significantly, ranging from high to low. Although the selection process was not entirely systematic, I acknowledge that it was somewhat accidental. Due to financial constraints, the study was geographically limited to the Kinki, Kyushu, and Kanto regions of Japan. Despite investigating only eight cases, I concluded that the objective of this chapter had been met, as the final interviewees demonstrated a tendency to refrain from attempting to enhance their clients' farming skills.

For each case, I conducted a semi-structured interview with the director or a staff member from the employment support service provider, typically responsible for instructing the laborers or service users. The key questions included: Why is the organization engaged in farming activities? What roles do the organization's director and clients play, and what expectations does the organization have for the laborers? Does the director attempt to improve the clients' farming skills? If not, how does the organization ensure the sustainability of the activity, particularly in terms of funding? These questions aimed to compare the cases in terms of the autonomy expected and allowed for the clients in farming activities. To clarify the concept of "autonomy," I introduced terms from the previous chapter, such as "operation," "observation," and "decision-making," as well as the collective expression "taking direction from crops." Additionally, staff members from four cases were invited as guest speakers at a symposium titled "Toward Creation of Agriculture-Related Jobs in Kamagasaki," held on December 19, 2019, in Osaka (Tsunashima 2021). Their remarks were incorporated as supplementary material to the interview results.

For two cases without a designated director, I visited the sites repeatedly to observe a variety of crops and other living things grown there. At each visit, when an interviewee focused on a specific life form, I initiated short, unstructured interviews to learn about their knowledge and techniques they had obtained or learned to grow it. The subsequent analysis aimed to connect these autonomous skill-acquisition processes with the same concepts used to understand the partnership cases. I also inquired about how the farming activities were sustained financially.

The approach outlined above is qualitative, as the interview questions included several terms that may have been unfamiliar to the interviewees. Had I not carefully chosen my words in accordance with their interpretations of my prior statements, the responses would have varied. This process allowed me to gather a broad range of perspectives on the agriculture-welfare partnership, providing a strong rationale for the chosen approach.

4.3 Results

4.3.1 Case A1: A Day Laborer

The above-mentioned 65-year-old day laborer (as of 2018) has been gardening independently in a suburb of Osaka since 2008. A carpenter since the age of 15 in Kamagasaki,[1] he saved his earnings to purchase a piece of land in the area (Fig. 4.1), though he was still unable to afford a house with a lifeline at that time. On this land, he maintains a kitchen garden, growing various vegetables, fruit trees, and ornamental plants, while also raising chickens and keeping bees. He only purchases rice and seasonings. Skilled in gardening, he propagates trees through grafting and layering, often sharing saplings with his neighbors. He originates from a landless branch of a farming family, and although he had occasional opportunities to help with farmwork, his knowledge of cultivation remains fragmented. When I asked him about the essentials for gardening, he unexpectedly emphasized the importance of observation. He stated, "Since no literature has shown me how to protect my trees, I have had to devise methods through trial and error. But now, I am too old to grasp everything by observation. It would be limitless if I tried to do so. You save time by asking someone else for advice instead." He sometimes borrows a smartphone from friends to search for information online and visits botanical gardens or urban parks to observe as many plants as possible. Despite the challenges of supporting and pruning fruit trees and managing pests, he maintains a strong desire for knowledge. He seemed to prefer working alone, although I attempted to involve him in a job creation initiative.

The facts gathered from field interviews are outlined below to illustrate how he accumulated knowledge independently. These descriptions are categorized by fruit tree, in the following order: orange (1), persimmon (2–7), fig (9, 10), avocado (11),

Fig. 4.1 Exterior view of A1's home garden

[1] A locality in Osaka City that was deliberately developed as a day laborers' town in the post-war period and gradually became known for having the highest concentration of homeless individuals in the country.

Japanese plum (12, 13), prune (14), cherry (15), chestnut (16), and those applicable to many species (17, 18), followed by vegetables and miscellaneous farm products (19–22). What is revealed is a range of functions that his observations serve, categorized as follows: The first category provides him with a compelling explanation for an inevitable phenomenon (type A; 1, 10, 12–15, 20). The second category offers a reason for the approach he took after the fact (type B; 2, 6, 7). The third category influences his decisions for future actions, sometimes providing new knowledge following experimentation (type C; 3, 9, 10, 21), or simply notifying him of something worth trying (type D; 4, 8, 11, 12, 14). This often relates to discoveries made off the farm. The fifth category drives him to specific actions out of necessity (type E; 5, 9, 16, 19). His knowledge accumulation also contains broader lessons that should be taught from a higher perspective (type F; 17, 18). Finally, he observes the work of other gardeners and models his practices accordingly to assimilate relevant knowledge (type G; 22).

1. "I bought a sapling of '*hassaku* orange' (*Citrus hassaku*) and planted it in my garden, but it was severely damaged by pest insects. This damage may have resulted from leaving pruned twigs piled near its base. Accepting the damage as irreparable, I left it alone for some time. However, the first fruit I found was less sour than expected. It is possible that the tree was actually Chinese citron (*Citrus natsudaidai*). Alternatively, the taste of citrus fruits may be more affected by maturation or soil conditions than by their inherent qualities. I find it fascinating to consider this possibility."
2. "The 'taishū-gaki' persimmon variety yields fruits that are prone to sunburn. Its saplings are expensive and grow more slowly than other varieties, but the astringent taste of the fruit can be removed quickly. The flesh is juicy before full maturity, but not as excessively sweet as that of 'fuyū-gaki,' even when fully ripe [author's note: He feared of consuming too much sugar]. Grafting joints are vulnerable to pest damage. Despite these challenges, I consumed a considerable amount of the fruit this year."
3. "'Taishū-gaki' trees bear both male and female flowers, making them useful as pollen parents, a fact not typically mentioned in the description. [He seemed to have learned through his observation this well-known fact that the plant species is monoecious.] When using this variety as a pollen parent, I plan to select a seed parent that produces sweeter fruits when cross-pollinated."
4. "I discovered an astringent persimmon tree planted in the grounds of a flophouse in Kamagasaki, but I could not identify its variety. I took a branch from the tree and grafted it onto another persimmon tree in my garden, and the branch produced larger fruits than those of common sweet persimmons. As a result, I now have trees of a different astringent variety in my garden."
5. "Persimmon trees extend thin branches that tend to grow too far and are easily broken. I always prune them with regret. Additionally, a steel pipe is needed as a prop, which lasts for about 10 years. The prop should be installed early to avoid damaging the roots. [If installed too late, it could harm the roots, which may have already extended beyond the point where the prop should be placed.]

Since persimmons produce flower buds at the ends of branches, the weight of the growing fruit can cause the branches to bend and touch the ground, leading to bruising. To prevent this, it is best to remove all branches except those growing from the trunk at least three meters above the ground. The root system tends to grow more slowly than the aboveground parts."

6. "I allow the harvest of persimmons to begin on September 1st, and I stop pesticide spraying on August 1st. As a result, a pest called the 'calyx worm' consumes the fruits, hiding behind the calyxes, which promotes earlier ripening."
7. "I have a persimmon tree that yields about 500 fruits. With proper care, I could increase this number to around 1000. If I had 10 trees, I could harvest 10,000 fruits. At a price of JPY 30 per fruit [far below the market rate] this would provide enough income for me. However, the cost of transporting the fruits within Osaka City would exceed the sales revenue, as I can currently only carry a limited amount."
8. "When the landlord of my additional leasehold garden allowed me to take some persimmon and orange saplings, I chose those with thicker trunks, expecting larger fruits. However, the stumps were too heavy for me to lift. To reduce their weight, I removed the dirt from the roots, which caused some damage. When transplanting them, I used bamboo props to stabilize the trees and prevent any vibrations from hindering the recovery of the damaged roots. At that time, the root systems had been reduced to a diameter of 60 cm. After transplanting, I placed water containers nearby to ensure the trees received adequate water. If water levels drop, the leaves become narrower."
9. Fig trees were covered with nets to prevent damage by boring cerambycids. "You should focus on thick branches. Thinner ones require less attention, as they can be pruned if damaged. In the past, to kill the insects tunneling through the stem and thick branches, I would poke at them with a piece of open copper wire. One day, I left the wire stuck in the tunnel, and the insects stopped expelling sawdust, which indicated they were no longer damaging the trees. I believe copper itself acts as an insecticide. Now, I protect the stems and thick branches in this manner, while I prune the damaged thinner branches after ensuring the fruit has ripened."
10. "In cities, fig trees can grow as large as desired, as there are fewer pests. However, in my area, they attract pests continuously. Therefore, I prune branches that grow too far apart and use them as cuttings, which delays the maturation of the fruit. This approach benefits my personal consumption but is not for sale. I also experimented by placing cuttings in water containers to promote fruit ripening, but the cuttings withered and failed."
11. "I recently saw an avocado tree bearing fruit at a park near Tennoji, just one stop by local train from Kamagasaki. I learned that avocados can also grow in Osaka, in addition to subtropical areas."
12. "Japanese plum (*Prunus salicina*) produces 1000 to 2000 fruits per tree. If a tree receives too much sunlight, its fruits ripen too quickly, resulting in a short harvest window. Furthermore, freezing the fruits diminishes their flavor. A neighbor's tree, which seems to enjoy a longer harvest period, is shaded for a

few hours each day during the season. I wonder if I could extend my harvest time by artificially shading my trees, based on closer observations of my neighbor's."

13. "Japanese plum trees appear to coexist with humans. They can only be independent after their roots have spread to a depth of 50 cm. Unlike persimmon trees, whose branches are prone to snapping, Japanese plum trees have supple branches, allowing me to shape them so that the fruit is within reach (Fig. 4.2). This shaping should be done after spring to ensure that any cracks from bending recover quickly. In autumn, the leaves are infested with caterpillars, but I leave them undisturbed and wait for the leaves to fall naturally."
14. "This year, I induce my prune trees to bear their first fruit. Prunes require close planting with other trees that flower at the same time to set fruits [due to their self-incompatibility]. In Miharu, a region where the name means the place where three vernal flowers bloom simultaneously, this occurs naturally. However, in warmer areas like Osaka, different prune varieties bloom at different times. I am curious about the method of preserving pollen, which I recently heard about."
15. "I failed to raise two cherry varieties, 'satō-nishiki' and 'napoleon.' Additionally, I could not cross these varieties because their flowering times do not overlap, and their pollen does not remain viable until later-blooming varieties are ready for pollination. The cultivar I am growing this year is the earliest blooming cherry variety, flowering in May, just before 'yoshino cherry' (*Prunus* × *yedoensis*). The trees I propagated by layering are now swaying in the wind, as I did not provide enough support. Bulbuls are the most troublesome pests for cherry fruit."
16. "I have been raising chestnut trees for 10 years. Initially, I planted a conventional variety and distributed the harvest to colleagues in Kamagasaki, but they soon discarded the chestnuts due to the difficulty of removing the astringent coat. Three years ago, I planted a new variety, 'porotan,' whose astringent coat is easier to remove. In the near future, I will need to prune the older tree, as chestnut trees typically weaken when they reach 20 years of age."

Fig. 4.2 Japanese plum tree with an artificially bended branch

17. "No literature has provided guidance on how to protect trees. As I age, I am eager to learn the results of my experiments."
18. "One should not entrust the care of fruit trees to professional landscape gardeners."
19. "I grow lotus for its edible roots in water containers. My observations reveal that the water level drops by 5–10 mm per day, allowing me to determine when to supply more water."
20. "I originally planted gingers in a sunny location, but I learned they should be grown in the shade. I created shade for the plants, but this led to root rot due to the sudden reduction in sunlight."
21. "As a child, I saw a white hen of 'shamo' [a Japanese indigenous breed typically with colored plumage] which sparked my interest in poultry farming. I have since hatched 60 chicks and butchered 20 chickens for personal consumption. I had hoped that the white hens would produce white chicks, and the 70th chick finally met this expectation. The cost of feeding the hens is ordinarily JPY 1500 per month, but it has jumped JPY 5000–6000 since I wanted a white chick. I observed that parent hens hold water in their mouths and dribble it onto their chicks to cool them during hot weather."
22. "A neighboring carpenter keeps honeybees, and I asked him about the dimensions of his beehives to make my own. In addition to a box made of 15-mm-thick planks, you need wire pieces stretched across wooden frames that fit into the box. This will allow easy separation of honey storage from the breeding areas in a hive. The honey storage can then be placed in a perforated container for filtration. To ensure the queen bee feels secure in the hive, you should introduce beeswax by March to alter the hive's scent."

4.3.2 Case A2: Autonomous Farmwork by a Former Trainee Receiving Employment Assistance

A participant in the action research project, described in previous chapters, began cultivating approximately 200 m^2 of farmland at his discretion in August 2017. Initially, he had been receiving employment assistance training since his 30s. At that time, he exhibited a particularly strong interest in vegetable cultivation compared to other participants. According to an interview, he had previously assisted his father in running a restaurant, where he was responsible for purchasing food materials, such as vegetables. This background likely contributed to his ability to excel at selecting fruits based on varying levels of maturity when harvesting, compared to others. His restaurant experience proved beneficial for his autonomy, though he initially appeared hesitant to take the initiative until I encouraged him to independently manage the farmland without receiving direct instructions. He commutes to the farm at least once a week, walking 20 min uphill from the nearest railway station, as he does not have a driver's license.

By the start of his small farm, using his own connections, he secured a location to sell his products weekly. His motivation for doing so stemmed from another vegetable grower, who had previously sold products in the same manner, wishing to expand the variety of vegetables in collaboration. This arrangement benefited both parties. Although he could have focused on fewer vegetable types for sale, he aimed to diversify his offerings to attract a wider range of regular customers through all seasons. They pinpointed his produce rather than happened to pass by. One customer even visited him specifically to buy mother tubers of taro, which are rarely available in ordinary stores. His maximum monthly sales reached nearly JPY 20,000, with an average of JPY 4500 for all products, or JPY 4400 and JPY 2400, respectively, when accounting only for the crops he grew himself (as of 2018). When attempting to grow a new crop, he often consulted the Internet for useful information, noting that the decision to grow it ultimately depended on his interest. For example, he had a desire to grow beets, and after learning how to cultivate them, he found they sold well. Additionally, he obtained new cultivation techniques via the Internet.

The following outlines the knowledge he gained during his first year of autonomous farming. Through observation, he enhanced his understanding of the relationship between crop characteristics and growing conditions (1–4) and confirmed information gathered from online sources (5, 6). Even when away from the farm, his work occupied his thoughts, and he reflected on his crop choices and strategies to increase sales (7–13).

1. He found a variety of turnips, having traditionally been cultivated in Nara Prefecture, growing unexpectedly slowly 3 weeks after sowing, and explained, "The colder it is, the slower the plant grows, with less risk of pest damage. Ultimately, it is a trade-off between faster growth and reduced damage."
2. Mulching the turnip with nonwoven fabric seemed to prevent cold temperatures from delaying its growth.
3. Lettuce and certain brassicaceous species appeared particularly susceptible to insect damage when direct-seeded. Consequently, he decided to grow seedlings indoors and cover them with nonwoven fabric even after transplanting.
4. Cherry tomatoes withered, seemingly due to summer heat, as it was impossible to maintain the plants with his once-a-week commute.
5. When stem lettuce began flowering, he sampled the entire stem. The lower part tasted bitter and tough, while the upper part did not. He decided to slice and dry the lower portion, as suggested online, while selling the upper portion raw.
6. He adopted a garlic variety called 'shishimaru,' explaining, "I make it a rule to plant small mother bulbs to harvest leaves, as instructed on a website. Medium-sized bulbs were planted in two rows on a ridge, while larger ones were planted in a row to ensure even-sized daughter bulbs. I weighed the mother bulbs to classify them accordingly."
7. On his way to the farm, he periodically visited a garden shop to select radish varieties for the coming winter. He chose varieties that would be suitable for leaf consumption, those with short roots that would swell easily, and those with reddish roots, which would be visually appealing when displayed.

8. For the current year, he selected the 'beni-azuma' variety of sweet potato, known to be suitable for baking.
9. Beet leaves sold well. Younger leaves were ideal for salads, while larger leaves could be fried. Naturally, mature plants produced beetroots.
10. Leaf mustard (*Brassica juncea* var. *integrifolia*) often sold out quickly at JPY 100 for five leaves per package.
11. Last winter, when cabbage became expensive, he sold Brussels sprout leaves (*Brassica oleracea* var. *gemmifera*), which were well-received.
12. During winter, when it was difficult to find crops to sell, he trialed two varieties of radish sprouts. Initially, he planned to sell them in a mixed-color combination, but the green variety grew too quickly, reaching marketable size in 10 days, while the red variety took longer. A package of seeds cost JPY 400, which he estimated would yield the equivalent of 20 packages.
13. Once, he attempted to sell parsley by cramming as many leaves as possible into each package. However, a customer remarked that it was too bulky to consume at once. He revised his approach, dividing the parsley into approximately 10 leaves per package, which proved more successful.

In 2019, he expanded his product line to include cherry tomatoes—which he had failed to harvest in 2018, as well as spice crops like stick fennel, dill, ginger, and Japanese ginger (*Zingiber mioga*), all of which I, as his former director, had not previously attempted to cultivate. Additionally, he developed several labor-saving techniques:

14. He learned to estimate the extent of the subterranean stem system of Japanese ginger by observing how far its flowers emerged from the leaves.
15. Knowing that ginger thrives in partial shade, he planted it alongside taro, which would provide shade during the summer. However, the majority of taro tubers rotted by May due to exceptionally hot weather. The remaining plants failed to extend their leaves, exposing the ginger to direct sunlight. Although he anticipated a poor ginger harvest, the plants ultimately fared as expected (Fig. 4.3).
16. After harvesting okra, with stems growing more than 2 m tall, he planted pea seedlings between the okra plants, allowing them to use the okra stems as support (Fig. 4.4). [Previously, I instructed participants to harvest bamboo from a nearby thicket and use it as support, a labor-intensive task.]

The descriptions provided can be categorized according to the same classification system used for Case A1: Type A (1, 4), B (2), C (6, 9–11, 13, 14, 16), D (5, 8, 12, 15), and E (3). The lack of Types F and G may merely be attributed to his shorter experience. His years of experience allowed him to evaluate the current year's crops in comparison to those of previous years, helping him identify how the climate had changed. As of 2019, he explained his current situation as follows:

> I have come to plan my work at my own pace. Even when I am off the farm, I often think about which vegetables to plant next. I always stay calm and never rush, which I consider my strength, though it might also be a weakness. I have received much help from my customers.

Fig. 4.3 Ginger harvested after an attempt at mixed cropping with taro

Fig. 4.4 Okra stems used as support for pea

4.3.3 Case B1: Support for the Former Homeless and Welfare Recipients

P, a retired 89-year-old farmer from Kyoto Prefecture (as of 2018), has provided guidance on farming to ex-homeless individuals and welfare recipients for nearly 30 years. Several homeless support organizations in Kyoto and Osaka referred homeless individuals to his farm, where he extended his house to allow them to live in. At its peak, P admitted up to ten people at once. Initially, he assessed whether they could endure staying in such a remote area for a month. If successful, he helped them register at his address, enabling them to obtain identification documents, such as a health insurance card, which would facilitate job hunting. Most of these individuals stayed for less than a year. A local welfare volunteer, commissioned by the municipality, sometimes assisted them with welfare benefits. Although P acknowledged that some required livelihood support, local residents often argued that outsiders should not receive welfare.

P observed that progress was easier when trainees showed a positive attitude toward farming. One trainee became a farmer in the locality, who had quit several years earlier, initially joined P's farm with determination. She would often tell P, "I have to grow better vegetables than you." Over time, she successfully grew vegetables and sold them to Japan Agricultural Cooperatives, the largest agricultural produce agency in the country. Additionally, she became highly skilled in making miso paste by sourcing materials economically, such as purchasing refuse soybeans, collecting broken rice, and obtaining salt by evaporating seawater with timber waste. Her miso paste became a popular product. However, despite her success, she struggled with managing money. If handled correctly, her earnings could have supported her financially. After her secondhand tractor broke down beyond repair, she changed her approach to work and life.

Since 2014, a local workers' cooperative has incorporated P's farming initiative into an employment assistance program for welfare recipients, funded by the municipal government. The program staff discovered that, despite initial difficulty in finding employment, the individuals could succeed when given appropriate tasks and the support of colleagues. Their motivation often stemmed from the desire to be helpful and to have a meaningful role. After completing the farmwork course, some trainees became more interested in agriculture and continued training at P's farm, while others improved their lifestyles and regained their appetite after physical exercise. Overall, completing the course fostered a sense of achievement and boosted their confidence. Around 2015, long-term trainees registered as producers at a local direct sales store opened by Kyoto Prefecture (Fig. 4.5), and profits from sales were distributed among them. However, the same issue reappeared.

> Each time they learned that their products had sold, they were happy. But later, they faltered. A year later, I suggested they save their earnings for the next year's production costs, but they repeatedly failed. I gave up trying. Now, I don't know how often I can intervene in how they spend money. Even with unlimited funds, I would be at a loss.
>
> Among the trainees, she was the only one who learned how to grow vegetables independently. The others were limited to tasks like harvesting, sowing seeds, ridging, and weeding.

It is evident that enabling trainees to perform a series of farm tasks autonomously is highly challenging. P acknowledged facing the same issue identified in my action research project. When I requested the interview, I mentioned that I hoped to gain insights into solving this problem. P responded: "We seem to face the same problem. It has been 30 years since I first took in homeless people, yet I still feel like I need someone to teach me how to address this issue." Despite this ongoing challenge, what motivates P to continue his efforts? He explained:

Fig. 4.5 Farm produce with trainees' names printed on labels laid out for sale at a local direct sales store

> What makes a significant difference is that the farm field is available to trainees. As long as you grow vegetables, you don't have to worry about losing your job—you can sell or consume what you produce. Moreover, farming is one of the easiest and quickest ways to regain a livelihood.
>
> One day, a municipal officer brought a welfare recipient to my farm. After working on the farm for a few hours, I was pleased to hear that the client had enjoyed the experience and expressed a desire to grow vegetables. The officer revealed that the client had previously spoken constantly about wanting to die. However, now the client was actively cultivating a field. Although still learning how to improve the quality of his produce, he now expressed a wish to become independent of welfare benefits. If he truly meant this, I believe he should change how he manages the money he earns. If he does so, my support for him will have been worthwhile.

P emphasized the importance of reconnecting people with the earth:

> It is essential to restore the relationship between people and other life forms. Today, people live far from the soil, surrounded by high-rise buildings and paved roads, and many perceive soil as dirty. When I ask trainees to do farm tasks without gloves, I explain that 'you will feel rejuvenated because the soil will energize you through the gaps under your fingernails.' For example, it is time-consuming but rewarding for them to sow vegetable seeds one by one into pots without gloves.

As P ages and his physical strength diminishes, his approach to training and his attitude toward trainees have evolved:

> It is useless to discipline those who rely on welfare with intense methods meant for athletes. Welfare recipients are more vulnerable to pressure, while athletes may benefit from such

> training. Forcing strenuous efforts can be counterproductive. In the past, I often wondered if they truly wanted to support themselves, even for a short time. In retrospect, I am glad I never expressed that. Now, I understand how difficult it is to balance rigor with tolerance. All I can do is help them engage with nature in a natural way, whether they are progressing or struggling. Know yourself, accept who you are, and find your own joy.

In line with this philosophy, the chief director of the employment assistance program set a goal not only to help clients secure regular employment but also to engage individuals who might withdraw from such opportunities by creating diverse job prospects. The cooperative has since expanded into food processing, in collaboration with local private companies. At the end of his symposium presentation, the director summarized P's philosophy on farming:

> Nature does not guarantee a consistent harvest, even with constant effort from growers. It is beyond human control. Farmers must employ various strategies to cope with this unpredictability. Unless they find joy in this challenge, they cannot sustain farming.

4.3.4 Case B2: Training Course Initially Designed to Support Homeless People

Q has been operating a non-profit organization since 2009 to bridge the gap between job-seeking individuals and the understaffed agricultural industry. The organization offers these individuals opportunities to regain their strength through vegetable cultivation, followed by assistance in securing employment with farmers or agricultural corporations. This initiative is structured as a training course consisting of weekly sessions over 6 months, culminating in agricultural job placement.

Applicants must complete an "entry sheet," a five-page questionnaire, and commit to recording what they have learned at the end of each session while attending all sessions punctually. These requirements present a significant barrier for those lacking commitment, as each individual may only take the course once. The course admits 6–10 trainees per term. Upon completing the course, trainees are required to fill out a form similar to the entry sheet, though their responses often differ significantly from those given at the time of admission. Q expects this shift to reflect the personal growth of the trainees. Past participants have remarked, "I am no longer so obsequious," "I have gained confidence in myself," and "I have become more tolerant of others," indicating a significant transformation in their self-perceptions and attitudes toward others.

Each session begins with an introduction, where instructors explain the farm task trainees will undertake that day. An instructor demonstrates the task first, and then participants follow. The organization emphasizes training trainees to perform tasks precisely as instructed, as the course is designed to prepare them for employment as salaried agricultural workers. The goal is not to enable trainees to work independently but to prevent the development of self-taught novices. Upon completing the course, trainees are ready to enter the job market, where instructors reinforce the principle of "When in Rome, do as the Romans do." The feedback from employers has generally been positive, with comments such as, "The new employee is diligent" and "They are reliable."

Throughout the training course, some trainees begin to take initiative beyond the given instructions. Q explains this phenomenon:

> It would be nearly impossible for them to do so if they focused only on the vegetables they were growing; they also paid attention to the soil, weeds, and wind. This is not always easy for those with a background in natural sciences, who excel at pattern recognition but may struggle with the tacit understanding required for holistic tasks. This elucidates the diversity in how people perceive and process information.

The organization has admitted a wide range of applicants, including one former trainee with high academic qualifications who had withdrawn from society for 10 years before enrolling. Other applicants have had borderline intellectual disabilities and struggled to work independently, often requiring someone to guide them. Homeless trainees, many of whom have construction experience, have been particularly adept at quickly understanding and performing farm tasks. Q suggests that a trend toward autonomous skill acquisition may emerge from the interactions between trainees from diverse backgrounds, for example, disabilities and homelessness. In this regard, she advises:

> The key to fostering autonomy is limiting instructor interventions as much as possible, allowing capable trainees to take charge, and praising their successful task completion.

Over time, the organization's approach to motivating trainees has evolved. Initially, trainees were paid for their work on the farm, as monetary compensation was more lucrative than other activities, such as collecting aluminum cans. This incentivized homeless individuals to participate enthusiastically, driven by their desire to escape homelessness. However, the organization no longer provides wages for attendance. This decision aims to refocus trainees on the purpose of the course, beyond monetary gain. At the start of the course, trainees are encouraged to spend 3 months on the farm without specific goals, allowing them to define their objectives individually. These goals may include earning income through farming, learning to grow vegetables, socializing with others, or finding a sense of belonging.

The organization has designed the course to be flexible rather than rigid, allowing staff to engage with each trainee on a one-to-one basis. If a trainee expresses genuine curiosity about agriculture, the course is well-suited to them; otherwise, it may be painful to trainees. Q notes that if a trainee is solely concerned with hourly wages, the organization will not recommend them for the course. However, if the trainee demonstrates curiosity, the organization places its trust in their potential. Q's approach is to support trainees' aspirations while refraining from interfering excessively. She avoids warning trainees about potential problems that may arise from acting against her guidance, maintaining a certain distance to manage any overly demanding behaviors.

Although the training course is not designed as a job creation program, there is a compelling reason to consider compensating trainees for their participation. Drawing a comparison with the financial industry, Q explains that while other industries factor human resource development into their production costs, agriculture has traditionally relied on unpaid apprenticeships. However, she cautions against compensating trainees for mere attendance. If wages were offered solely for following instructions, the situation would resemble that of convenience store work, where employees simply follow managerial directives. This, however, does not apply to agriculture, where rewards are linked to the results of one's labor, not the number of hours worked. The course aims to teach trainees that success is determined by the quality of their efforts, and that true belonging is achieved through initiative and personal responsibility.

The course primarily targets individuals aspiring to become salaried agricultural workers. These workers are expected to be punctual and compensated for their labor without assuming any significant risks. However, some trainees exhibit entrepreneurial tendencies, as they choose to operate at their own pace, despite the associated risks. Over time, the staff increasingly differentiate between trainees suited for salaried positions and those inclined toward entrepreneurship. Some of the homeless individuals known to the staff were once entrepreneurs or company owners; for them, following another's instructions can be difficult. It is also possible to remain homeless while growing and selling vegetables. An alum of the program, who was once a recluse, advanced to an agricultural academy and eventually became a farm manager.

During each session, the instructor works to maintain a balanced atmosphere of mild tension. At times, laziness can spread, creating a relaxed environment, especially when trainees show little concern for how they are perceived by others. Some trainees may also influence others negatively. For instance, former recluses who are highly committed to the program may bring excessive perfectionism, which can impact the group. In such instances, the staff must take action to restore balance between tension and relaxation. The instructor varies their tone of voice depending on the situation. Other strategies for maintaining the atmosphere include involving younger staff to encourage trainees to regain their sense of honor, offering enthusiastic encouragement, moving the worksite, declaring a break, changing tasks, or grouping trainees differently. These techniques, honed through years of experience, help the staff maintain an effective learning environment.

The organization employs a variety of strategies to teach agricultural techniques. For example, when explaining the importance of deep furrows and high ridges in tomato cultivation, the information may be relevant to some trainees with higher academic qualifications, but unnecessary for others. This contrast often arises between former recluses and homeless trainees. Staff members must carefully assess whether their explanations should focus on practical, surface-level advice or more comprehensive theoretical concepts. In either case, the staff prioritizes ensuring that trainees understand the fundamental principles of farming. To achieve this, the organization provides spaces where trainees can freely ask questions without fear or hesitation.

Mistakes occur not only in maintaining the appropriate atmosphere but also in vegetable cultivation. However, these errors provide valuable learning opportunities, often leading to successful comebacks. Staff members explain the causes of failures and the measures they have implemented to address them. Maintaining strong interpersonal relationships is essential for ongoing learning. It typically takes 3 months or more for staff members to fully understand each trainee's capabilities. Only after this period can the organization confidently recommend trainees to potential employers. The following examples illustrate successful cases of former recluses who reintegrated into society: First, one alum, who had social phobia and struggled with communication, secured a position as a stockbreeder. In this role, he interacts primarily with his immediate superior, and his colleagues describe him as overly sensitive but highly dependable, as he can perceive even the slightest changes in the cows' conditions. Second, another individual, sharing a similar personality, found employment at a family-run farm. The farm manager, also averse to social interaction, communicates minimally with his subordinates. This unusual combination has led to mutual understanding without verbal communication, which the manager considers invaluable.

As of 2018, the organization has supported 13 municipal governments in implementing the Independent Support Program for the Needy. One former trainee, employed in a cleaning service, took one day off each week to participate in the program. The training site became a space where he could express his frustrations about his new job and maintain emotional stability. He once told a staff member that it was the only place where he could converse while performing menial tasks, such as mowing weeds. Additionally, the organization manages allotment gardens for 150 families or groups and offers training sessions for companies to provide in-house education for new employees. These services are funded by client fees, which help support the admission of needy applicants into the training programs. Whenever possible, Q emphasizes the value of learning to grow vegetables, stating, "To grow vegetables is to grow yourself, and it is worth learning how to do so under any circumstances."

The organization is also involved in vegetable production. From this perspective, Q asserts that the current agriculture-welfare partnership still requires a mutual understanding between the two sectors, which have often struggled to align their objectives. The agricultural sector seeks laborers who will contribute to farm management focused on optimum crop production, while the welfare sector needs to provide job opportunities for service users. The primary difference between these sectors is their approach to the time spent on client guidance: the agricultural sector views this time as a loss, while the welfare sector does not. Despite this, many farmers provide guidance without compensation. It is clear that the partnership will be short-lived unless both sectors collaborate to modify their approach.

4.3.5 *Case B3: Autonomous Farmwork by a User of Employment Support Service for Disabled People*

Since its establishment in 2014, a sheltered workshop in Osaka Prefecture has provided individuals with mental, intellectual, physical, or developmental disabilities the opportunity to engage in farmwork. The initiative began when a neighboring

Fig. 4.6 A client working independently from staff instructions

farmer requested assistance with labor, as the workshop staff were reluctant to undertake monotonous, light tasks. As a result, clients showed a significantly higher attendance rate, accompanied by relaxed expressions and animated conversations. The staff soon realized that the most effective counselors for these clients were the sunlight and soil they worked with. Consequently, the workshop expanded its activities to include tasks such as handyperson services, house cleaning, and the washing of welfare products, although one client's case deviated from these trends.

As of 2018, farming activities accounted for only 1% of the workshop's total revenue but continue to serve as a marketing tool linking the workshop with local businesses. The workshop has since leased a smaller plot of farmland from a neighboring farmer. Using the "Type-A continuous employment support" program,[2] one client commuted to the land parcel every weekday (Fig. 4.6). While staff members designed a seasonal cropping plan for her, she frequently altered it. She also refused to thin seedlings, explaining, "I feel sorry for the sprouts." More notably, she sowed and planted seeds and seedlings provided by neighboring farmers. It appeared that she carefully observed the crops growing in nearby fields to determine the optimal time for sowing or planting. Staff described her as a professional at seed production and germination. Although she struggled to articulate her cropping plan, it seemed she had her own approach. Consequently, the staff's role was to gather the necessary information to help her execute her plans, such as the process for blanching Welsh onions.

[2]The Ministry of Health, Labor, and Welfare targets people with disabilities who are currently unable to obtain "regular employment" by implementing the following support schemes based on the Services and Support for Persons with Disabilities Act enacted in 2006: (1) "employment transition support" for those seeking regular employment, and (2) "continuous employment support" for those with difficulty in obtaining regular employment. The latter is further subdivided into Types A and B. Type A requires an employment contract between the service provider and the user and is subject to labor laws, including the minimum wage system, while Type B is not. In addition to reorganizing conventional sheltered workshops, new enterprises have emerged to administer these schemes.

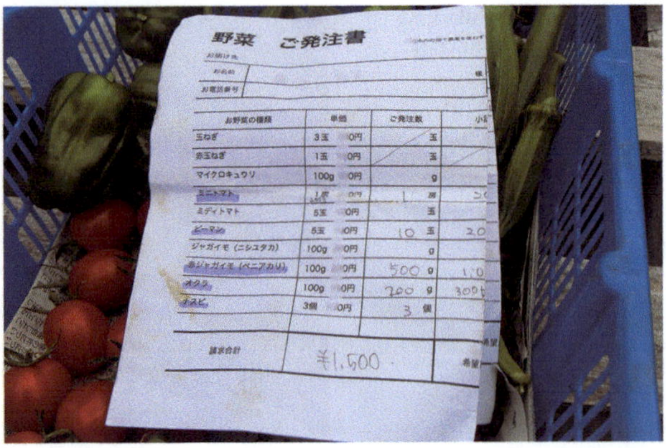

Fig. 4.7 Harvested vegetables and a delivery slip

To turn her gardening activities into a revenue stream for the workshop, staff members also attempted to sell her produce directly or through a delivery service. Due to her higher cerebral dysfunction, the staff handled tasks like weighing and pricing the products (Fig. 4.7). Despite these efforts, her gardening activities had not yet generated a profit, and the municipal supervisory authority expressed skepticism. However, she asserted that working the soil had significantly changed her from what she had been at previous indoor work in another facility.

The following examples illustrate the client's autonomous approach to gardening, which often exceeded staff expectations, and the assistance she received:

1. Inside the greenhouse, sweet corn grew. As the client refrained from thinning, the plants became overcrowded with unproductive growth and reached the ceiling.
2. Staff members believed that a neighboring gardener had provided the client with more okra seedlings than could fit in the allotted space. Consequently, she planted them along walkways.
3. In the local area, farmers harvest burdock leaves and young roots before planting soybeans for green beans. The client appears to have modeled her practices after these local farmers.
4. A staff member intended to transplant Welsh onion seedlings after drying and separating their roots. However, the client had already planted them, causing overcrowding.
5. After all clients left for the day, staff members delivered farm produce to several delicatessen shops.
6. Since client service hours are fixed, gardening activities must continue during the day, even in hot weather.
7. Although the staff may request help from other clients to control rampant weeds, the organization prioritizes more profitable activities over gardening.
8. The staff hopes the client can eventually start farming or secure employment as an agricultural worker, though her difficulty in cooperating and communicating with others remains a significant barrier.

4.3.6 *Case B4: A Center for Type-B Continuous Employment Support*

The organization operates a center for "Type-B continuous employment support," as defined by the Services and Supports for Persons with Disabilities Act. Its mission is to assist working-age individuals in their social reintegration and help the elderly maintain community involvement or develop a sense of belonging. Initially, the center could only pay clients JPY 100 per hour for their work. Since opening a farm, this rate has increased to JPY 300 or 400. Clients are referred to the center by psychiatrists, with supervisory authorities recommending that some clients engage in physical exercise or outdoor occupations, although not specifically farming. By opening its own farm, the organization has expanded the range of clients it can accommodate. Clients vary widely in their abilities; some can manage piecework, while others require a certain distance from their peers.

Currently, 40% of clients participate in farmwork regularly, with an average of 12 clients per day, a maximum of 15, and a minimum of five. Staff have noted that these clients seek physical exercise and strength-building. However, a 20-min walk from the nearest train station to the farm can deter some clients, and hot weather further reduces participation during the summer. When clients cancel at the last minute, the organization typically puts extra staff or volunteers on standby.

The director assigns most clients light tasks such as harvesting and packing vegetables, while staff and local volunteers manage heavier tasks like ridge-making and weed mowing. The organization must be prepared for situations in which light tasks prove challenging. For example, one client picked underdeveloped okra fruits, teaching staff that instructions need to be more precise, such as specifying the required fruit length. Clients with autism sometimes perform farm tasks with excessive precision, which can reduce time efficiency from a labor-saving perspective. While such matters would not be considered in a purely profit-driven business, they are significant for the organization's mission. According to the director, the success of the agricultural-welfare partnership depends on tailoring job opportunities to individual clients. If clients are merely assigned tasks dictated by the support center, agriculture becomes no different from any other occupation.

The center generally does not offer guidance for more technical farm tasks, such as using hoes or controlling cultivators. Most clients view farmwork as one of several occupational options, although a few express a desire to become farmers. The organization allows these clients to gain experience in specialized tasks like pruning or picking. One such client recently set a goal to find employment as a farmer or florist after 3 years at the center. Clients with an interest in agriculture often observe their surroundings for other tasks they might complete beyond their assigned duties. Some even attempt to cook with the produce they bring home. The organization encourages clients to eat what they have grown, a practice the director refers to as "loose gardening."

The organization emphasizes the importance of social connections between clients and the local community. Its primary goal is to help clients lead satisfying communal lives. Clients' quality of life improves when the vegetables they grow contribute to local diets or become part of product lines for companies, with feedback from consumers. For instance, clients learned that their vegetables were used in salads sold at shops near the station. Another consignee included photos of dried Chinese radish strips in its catalogs. The consignee, involved in the "*teikei* movement", also organizes come-together events where consumers can meet the producers, including the clients. Additionally, the center provides "canteen for children" food materials, such as irregular-shaped cucumbers or worm-eaten greens, which are safe to eat and even tasty but unsuitable for sale. Staff members communicate any positive feedback from consumers, such as reports that clients' products were enjoyed, to the clients.

For clients, the significance of such feedback far outweighs concerns about wages. The organization continues to advocate for these positive responses. While the support center also handles non-agricultural contracts, it is difficult for clients to visualize how their produce reaches consumers. Staff understand the need to simplify this process, and they believe agriculture offers clear advantages in fostering a sense of connection between clients and consumers.

The director encourages clients to consume food in season, emphasizing its healing potential over medication. The organization uses a nearby group home to allow clients to share meals, revealing that elderly clients from farming households hold distinct views on food. Staff are confident that interactions with farmers or food industry workers can improve clients' lives, as some have been inspired to cook for themselves. Some clients now prioritize locally produced or handmade food, while others have begun to experiment with making confections at home after learning that similar items sold well at the center.

The primary criterion for crop selection is the need for labor during the off-season, along with marketability. Chinese radish is a good example, as it is harvested in the latter half of winter and processed into dried strips by hand. This crop offers clients an opportunity to develop slicing skills. The organization sometimes faces shortages of products due to its preference for outdoor cultivation, which follows the natural growing season. To offset seasonal dips in sales, the organization processes produce, though this involves more investment in equipment and is less profitable than selling fresh vegetables. Combining food processing with farming allows for a more flexible schedule.

In addition to Chinese radish, staff grow tomatoes and eggplants, which are in steady demand from local regular customers, despite considerable technical skills involved in cultivation. Face-to-face sales require a variety of vegetables, including unique or appealing options. Sweet potatoes are planted with other crops in the summer, as they require minimal attention during peak farming season. In the past, the producers grew the common variety 'naruto-kintoki,' but later found it difficult to process. As a result, they switched to 'beni-haruka,' a variety with a finer texture that lasts longer and offers more processing options, including baking for direct sales and drying after steaming for retail. The most notable crop selection is Malta jute (*Corchorus olitorius*), for which the organization has developed a method to involve as many clients as possible in its commercialization. After pruning, clients

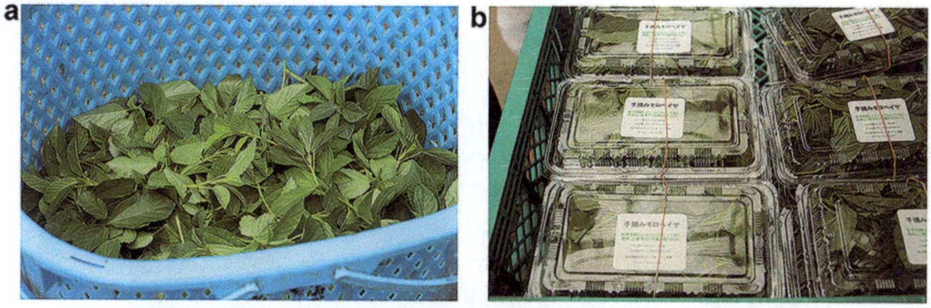

Fig. 4.8 The main product in summer season mentioned in the interview: (**a**) Harvested shoots, (**b**) Ready-for-sale packages

pick new shoots and pack them, and the product can be stored in refrigerators for a week (Fig. 4.8). To mitigate the risk of crop failure, the organization grows Malta jute both outdoors and in greenhouses, ensuring high quality and reliable harvesting, even after rain.

4.3.7 Case B5: An Employment Support Center for Disabled People Derived from an Organic Farmers' Commune

The founder reclaimed forests on his parental property in the late 1980s, where he established an organic farming commune. This commune aimed to create a self-sustaining community for people with disabilities, distancing itself from the conventional social welfare system. After learning that it was difficult for beginners to make a living solely through organic vegetable production, the founder expanded into poultry farming. He assigned disabled members tasks like egg collection and weeding, though their abilities were limited in some areas. To compensate, the founder personally delivered farm products to regular customers and managed additional tasks, such as tending Welsh onions and spinach.

As the commune grew, the founder adapted the property, transforming it from a small sheltered workshop into a social welfare corporation. The organization now houses centers for "Type-B continuous employment support," "employment transition support," and "daily nursing care," as outlined in the Services and Supports for Persons with Disabilities Act. Applicants are referred by municipal offices, hospitals, and operators of the "consultation support" business, with the latter being particularly active in referrals. Some applicants contact the organization directly to inquire about participation in farming activities.

Staff members hold a daily morning meeting to assign tasks for the day, emphasizing the importance of gaining client consent. Clients with mental disorders, serving as associate staff members, contribute suggestions on task execution, which full staff members respect. After the meeting, the group travels by car to the poultry farm, a 15-min drive away.

Fig. 4.9 Inside of a hen house equipped with feeders and egg-laying boxes

The farm's henhouses (Fig. 4.9), which house 100 chickens each, require eggs to be collected three times daily. Clients check nest boxes, and staff inspect the surroundings for eggs that may have been laid outside[3]. In one incident, a client mistakenly mixed some laid on an earthen floor, which were actually fertilized, with those for delivery, leading to a minor problem since the eggs were to be delivered to the organization's restaurant. When the chickens reach 15 months of age, they are slaughtered by a professional subcontractor and processed into sausages served in the restaurant. Two staff members are regularly assigned to farming tasks, including building henhouses and creating boxes for egg-laying and chick-raising, all designed to prevent hens from laying eggs on the ground.

The repetitive nature of poultry farming allows the director to plan each client's role in advance, such as assigning specific henhouses for egg collection. Staff supervise small groups of clients, with a particular focus on those who are more "sluggish" or prone to panic attacks, for whom a screen or net is used to provide a calming environment. Recent staff members have streamlined the egg-encasing process to simplify tasks and involve as many clients as possible (Fig. 4.10). As a result, tasks that used to take a staff member 30 min are now completed by clients in 1 h. Before the introduction of cushioning materials like corrugated or newsprint paper, clients frequently broke up to 10 eggs daily; now, such incidents are rare, and only one complaint about cracked eggs has been received since the change. One client, once considered "the poorest student" in special education, has since developed sufficient skill to handle eggs without damage. Staff also deliver eggs to customers, including users of Daily Life Care services. One middle-aged user, previously withdrawn, has

[3]The eggs found on the floor imply that they were laid some time ago, making it impossible for growers to determine their "consume by" dates.

Fig. 4.10 Encasing of eggs by clients

shown noticeable change, possibly due to his newfound interest in chickens, which highlights their strong presence in the center.

Crops such as soybeans, sesame, barley, and foxtail millet provide clients with tangible evidence of their labor, motivating them to remain engaged. As the organization has no combine harvester, harvesting and threshing these crops requires collective effort, and once completed, clients appreciate the scale of their work as staff demonstrate the quantity of grain handled. The founder observed that clients seem to enjoy tasks like harvesting and weeding without hesitation. The crop field also grows garlic, Japanese scallions, and burdock, while the orchard contains blueberry and plum trees. Land preparation is the responsibility of the staff. Blueberry harvesting takes place from June to September, with staff and clients working together to complete the task efficiently, though the yield may be somewhat coarse. Impurities can be removed later. The harvested blueberries are sorted, with the unsuitable fruit being frozen and processed into jam.

The organization sells its produce twice a month at a nearby cooperative store. Staff and clients collectively harvest and package the goods. At one point, the organization was allocated a sales booth quota for social welfare enterprises at no cost. However, competition from the organization's produce led to complaints from the store, and the founder subsequently opted to pay for a quota intended for general enterprises, which required 15% of total sales figures as venue fees.

A limited number of clients, particularly those with autism, are involved in food processing, where tasks are clearly outlined each morning. In addition to its farm products, the organization receives substandard food materials—such as ginger tubers, Japanese scallions, taro, and sweet potatoes—from an organic farmers' association based in a neighboring prefecture. These materials are dried or made into pastes. Tasks like peeling sweet potatoes and garlic can be shared among any clients. After someone discovering that soaking garlic in water makes it easier to peel, all clients follow this method. The organization employs a specialist in baby food processing who oversees production and supervises some clients. A subsidy enabled the purchase of an autoclave and a quick freezer, the latter crucial for maintaining the quality of grated radish for more than a year.

Feedback from consumers and consignees has been frequent and encouraging. Clients are motivated by such feedback, which reassures them about the quality of their work. Staff members emphasize the importance of hand-peeling food materials and carefully screening unpolished rice, demonstrating the significance of each production stage. The organization values the time spent on communication between staff and clients, as it contributes to a greater understanding of the manufacturing process.

The organization has enabled the payment of labor wages on an hourly basis by dispatching clients to external workplaces. Clients who are likely to deviate from staff expectations, such as those who go somewhere with a weed they have just removed, will behave differently in these outside settings. A large-scale agricultural production corporation hosts a party to express appreciation for clients' services, particularly when they have worked alongside local elderly individuals. Initially, the founder believed that commercial businesses would hire clients with disabilities, assuming they would be inexpensive and compliant labor. In the current locality, it would be inappropriate to demand that employers ensure clients receive at least the legal minimum wage. For example, when the founder negotiated a contract with a local meatpacker to process pork into miso paste, the partner claimed the company could only afford to pay for the labor of three part-timers, which they considered sufficient for the task. Despite agreeing to this lower payment reluctantly, the organization assigned the work to a team of six workers, including both staff members and clients. Later, the agricultural production corporation applied for a subsidy to install machinery that would make it easier for workers with disabilities to perform their tasks. Consequently, the founder began teaching local businesses how to apply for such subsidies through the municipal social welfare authority.

The commune, the precursor to the current support center, had space for no more than ten clients at a time to foster mutual aid. As the organization transitioned from a commune to a sheltered workshop and later to a social welfare corporation, it faced increasing demand to accommodate slower clients. Some clients require two staff members for supervision. If the organization had only accepted applicants who were highly motivated to work, the staff could have provided strict training upon admission, with the clients' consent. While accepting slower clients risks productivity, there is a strong taboo against refusing qualified applicants even when vacancies exist, simply due to low motivation. Meanwhile, supervisory authorities are advocating for higher labor wages for clients, and prospective applicants increasingly prefer organizations that pay a certain amount. One potential solution to this ambivalence, employed by several other social welfare corporations, is to separate more motivated clients into employment support programs and leave less motivated individuals in the existing sheltered day workshops. However, the founder rejected this approach, which is why the organization recently began cultivating some cereal crops to double the labor charges paid to clients. The founder fears that failing to adopt the "Double the Labor Charge" slogan could impact service fees paid from the municipality, which are crucial to the organization's revenue. Meanwhile, these cereals are prone to rainy damage in the harvesting season. Thanks to the clients' exceptional focus, they and the staff were able to complete the harvest of a 30-are farmland with sesame and lodged foxtail millet, before a typhoon struck the farm.

4.3.8 Case B6: An Employment Support Center for Disabled People Derived from Local Connections Centered Around an Organic Farmers' Association

The employment support center originated from a producers' cooperative association oriented toward organic farming. This initiative was facilitated by the Japan Organic Agriculture Association, which organized a meeting in the prefecture in 1974, providing an opportunity for its formation. Since 2018, the center has been managed by R, the third-generation chairperson of the association. The association's primary product is lotus roots, the cultivation of which dates back to the post-Russo-Japanese War depression. At that time, R's grandfather, a member of the local judicial agricultural union, introduced lotus cultivation from an island in the Inland Sea of Japan.

While the nationwide Rice Yield Boosting Movement began in 1956, a policy shift in 1971 promoted reductions in rice acreage. Farmers were assigned specific reductions in acreage. The association, securing agreements with several village communities, took on these reductions. This allowed those villages to maintain the acreage under rice cultivation, while the association planted lotus in their paddy fields of the same acreage. The association began marketing this novel produce, which is harvested between late July and October (Fig. 4.11). After harvesting, the cold winter months preserve the unreaped lotus roots in a dormant state, similar to refrigeration. Thus, producers only harvest on orders from consignees or customers, and the crop requires less labor than fruit vegetables like cucumbers, which require harvesting twice daily during the summer.

Initially, R decided to grow lotus without pesticides. However, given that farmers typically used herbicides, and insecticides as well because petioles would be broken to decrease yields once aphids damaged the crop in its initial stage, R encountered challenges. A local consumer cooperative, with which the association had a trading relationship since 1976, requested that the association market lotus roots with the reduced amount of spray at least, on condition that pesticide-free cultivation is deemed unrealistic. The association agreed to this arrangement.

Since then, the association members have refrained from using pesticides except for designated insecticides. Instead of herbicides, they use a power-driven mower between rows and hand-weed between roots, which both churns the water and

Fig. 4.11 Harvesting of lotus roots

shades the bottom to prevent weed seed germination. The association permits neither herbicide spray even on levees. Not being permitted to apply any chemical fertilizers, members pursue the best taste that they can give with only manures.

Effective insect pest management requires diligent observation. The association's policy is to promptly address pest damage by spraying insecticides one or two times at an early stage of infestation. Delayed applications allow some pests to escape and increase damage. After the introduction of pheromone traps, following guidance from the municipal government, the association ceased using insecticides for tobacco caterpillars (*Spodoptera litura*). Successive chairpersons of the association have prioritized building strong relationships with consumers, recognizing that mutual support between producers and consumers is essential for securing both safe food and long-term sustainability. The association learned the importance of maintaining consumer relationships after watching neighboring villages stop lotus cultivation.

In line with a 1984 system providing vocational training for disabled people, the predecessor of the current center employed graduates from a local school for disabled children, whose principal was R's cousin. In the early 1990s, the prefectural government established a sheltered workshop in the same locality. Consequently, two clients were sent to R's farm daily. Initially, some staff members believed farmwork would be monotonous, but soon they observed that the clients returned to the office with "the happiest smiles on their faces." Today, R's farm is recognized as a Type-B continuous employment support center, exempt from minimum wage requirements. The center also employs a laborer with a disability, on a referral from the local Public Employment Security Office. R ensures the laborer is exceptionally paid the minimum wage, appreciating their hard work, though noting that mistakes can occur when excessive trust is placed in the worker. For instance, a batch of rice husks to be charred was completely burned due to improper instructions.

Fig. 4.12 Manure material and product ready to ship

The laborer and clients, all of whom have no prior farmwork experience, are responsible for various tasks, including weeding, vegetable harvesting, and manure preparation. The center ships approximately 2000 bags of the manure product annually, each weighing 15 or 20 kg (Fig. 4.12). The materials—chicken droppings, soybean cake, and rice bran—are procured from local producers and mixed. The chicken droppings are supplied by farmers who send their eggs to the aforementioned consumer cooperative, after passing quality inspections. These producers transport the droppings to the center's farm via dump trucks. The soybean cake is supplied by a producer of fermented soybeans (*natto*), who previously provided excess materials to a now-defunct livestock farm. The center became the alternative partner for this producer. Substandard soybeans in the cake are used as green manure, contributing to the center's cyclical resource use and mutual assistance.

When harvesting garlic and onions, the farm receives support from several employment support facilities for disabled people. Although participation is mandatory, some work actively while others are reluctant. R acknowledges the varying abilities of participants but insists that even in an "agriculture-welfare partnership," some financial returns must be generated. When faced with slow workers, he offers them alternative tasks, though he could refuse them entry. Fortunately, the two current clients in the Type-B program seem to enjoy their work. R plans to plant onions on the farm's 6000 m^2 area with the clients.

R is also a board member of a social welfare corporation, where he is the only member with an agricultural background; the others are involved in social welfare or education. He has realized that his approach to work differs from that of the other board members. This experience impressed on him that he was dependent on the help of others, particularly people with disabilities. As he concludes, "Bringing out the best in them enhances my well-being because they help me avoid situations where I would have to work alone."

4.3.9 Case B7: An Employment Support Center for Disabled People Juxtaposed with Group Homes in a Plain Area

In 1993, in response to the decline of the domestic sewing industry, the director's parents shifted to social welfare services, converting their home to accommodate disabled individuals. At the time, there were limited options for disabled people to live communally and work for wages. In 1996, the family incorporated the enterprise with the mission of "creating an environment where disabled and other disadvantaged individuals can lead independent, communal lives according to their discretion." The director recognized that while disabled individuals had the potential to be versatile workers, they were often relegated to underpaid subcontracted labor within the capitalist economy. Consequently, the organization decided to focus on agriculture, believing that, once united, the clients could serve as a vital part of the agricultural workforce. The director also believed that unemployment was virtually nonexistent in the food production sector. The introduction of the Services and Supports for Persons with Disabilities Act in 2006 further influenced the decision to switch to farming.

The director asserts that personal preferences should not influence social welfare services. As such, the corporation serves not only individuals with physical, intellectual, and mental disabilities, but also long-term unemployed youth, withdrawn individuals, single mothers, survivors of child abuse, orphaned adults, individuals with chronic illnesses or early-onset dementia, and former convicts. New clients are typically referred by local welfare offices when other providers find a case particularly challenging, or when individuals express a desire for admission. Upon receiving a new client, staff initially provide day care and evaluate whether to offer employment or apply for Type-B continuous employment support. If a client does not possess a disability certificate, the staff assists them in obtaining one before initiating any welfare services.

As a social enterprise, the corporation collaborates with various local stakeholders. First, it sells onion seedlings to professional farmers and receives guidance on crop cultivation, as well as farmwork contracts. Second, the corporation provides vegetable seedlings and technical support to sheltered workshops for disabled individuals, purchasing substandard produce from these workshops. Third, it assigns clients who prefer not to engage in farmwork to jobs with commercial companies seeking to meet legal employment quotas for disabled individuals. In this capacity, the director also acts as a consultant for several of these local businesses. Additionally, the corporation trades farm produce with these companies and frequently advises the central government on unprecedented welfare cases in exchange for public funding.

In 2004, the corporation's farm began with a small vinyl greenhouse, as staff struggled to find work for clients on rainy days. They initially attempted to grow mesclun, but the effort ended in failure. Subsequently, they sought guidance from a local senior farmer, resulting in the packaging of the farmer's produce. Later, the staff and farmer focused on growing apios (*Apios americana*), successfully propagating the tubers, though they had no market value locally. Through numerous failures, the director discovered a viable path forward. He realized that clients could produce onion seedlings, even though they struggled to mature them. The director decided to focus the organization's efforts on tasks that clients could perform independently, reflecting his understanding of the diverse needs of the clientele. For instance, a client with an intellectual disability is assigned to the hydroponic section, where repetitive tasks are tolerable, while more skilled individuals are given tasks that change daily. Clients who frequently complain about coworkers are given tasks they can perform alone or facing the same direction. Jigs are adapted to accommodate the clients' physical needs. Importantly, the director acknowledges each client's current abilities, such as a client with a driver's license and severe intellectual disability, who is tasked with delivering farm products to retailers, or another client who, unable to find employment in typical companies, has displayed exceptional computer programming skills by designing a mowing robot.

As of 2019, the corporation operates hydroponic greenhouses covering 2000 m^2 (Fig. 4.13), a 2-hectare olive orchard, and 4 hectares of outdoor upland fields. Most of the farmland is leased from local professional farmers who have been confronted with labor shortages. The client group manages the entire process of onion cultivation, from sowing to shipping, with staff only determining the dates for planting, transplanting, and harvesting (Fig. 4.14). Chinese cabbage (*Brassica campestris* var. *amplexicaulis*) is also grown but requires professional assistance during harvesting. Both crops are sturdy and marketable. In contrast, other leafy vegetables are too perishable for clients to handle independently and require supervision, while root and tuber crops, such as potatoes, are often unmarketable and result in financial loss.

The primary challenge in the "partnership between agriculture and social welfare" is that social welfare organizations practicing agriculture have yet to achieve full integration into the agricultural industry. The director insisted that,

Fig. 4.13 Inside of a hydroponic greenhouse

Fig. 4.14 Onions ready to ship

as a public organ, service providers for disabled individuals should address this challenge by fostering collaboration with local professional farmers and businesses. When a welfare organization lacks certain resources, neighboring agricultural entrepreneurs may offer both the required materials and guidance, ultimately entrusting related tasks to the disabled workers. In this way, once organized into teams, individuals with disabilities can divide labor to compensate for one another's weaknesses, functioning as integral contributors to the food production industry.

For various reasons, including aging, clients will become unable to work. In such cases, they can receive assistance from animals and technological innovations. For example, when a client with presenile dementia feels compelled to wander, staff may allow them to keep a goat on a leash. The goat will graze on the weeds along their chosen path, effectively performing a farm task. Additionally, the organization has introduced a type of hive that enables beekeepers to easily collect honey. These innovations reduce unnecessary physical labor for clients, giving them more time for creative activities and socializing during teatime. The director emphasized that it is unjust to assign workers with disabilities to monotonous but easy tasks, such as weeding, solely because of their disabilities. The goal is for individuals to contribute to society to the best of their abilities, working together to build a humane community where everyone can take on a meaningful role.

In conclusion, the director underscored the distinction between farming and other types of work in the context of Japan's recent "work-life balance" movement. Achieving this balance is challenging for most people, especially for clients with mental or intellectual disabilities. While society has artificially separated work and life, true balance begins with recognizing the unity of these elements. In farming, one cannot defy the laws of nature. Attempts to force this balance may lead to an over-reliance on pesticides, which harms the soil. Unlike urban work, which often undermines both physical and mental health, farmwork, while physically tiring, can be mentally restorative. Ultimately, it may provide a sense of relaxation more assuredly than idleness, offering a reunion of work and life.

4.3.10 Case B8: An Employment Support Center for Disabled People Juxtaposed with Group Homes in a Mountainous Area

The current director of the farm shared the concept of a "partnership between agriculture and welfare" with a nurse who was raising goats while providing home care services at a regional hub city near the farm. They believed that goats, being friendly animals, would not intimidate people with disabilities (Fig. 4.15). Goat rearing was a novel activity in the locality, which further motivated them. In 2005, the director incorporated the farm to sell its dairy products. Initially, while advocating for collaboration between agriculture and social welfare, the director had no clear vision for how to implement this idea. Beyond the goat farm producing dairy, the organization also manages 1-hectare paddy fields, a 50-are upland field, a half-year-old chestnut grove, a chicken farm, and a greenhouse.

Around 2013, the director's business partners established a sheltered day workshop within commuting distance from the farm to serve as centers for Type-B continuous employment support and employment transition support. The farm began offering job opportunities to clients of these centers, and a former client joined the farm staff, now overseeing vegetable production. Initially, the center barely managed to pay both staff and clients a minimum wage, which became unsustainable after 4 years. Consequently, the center applied for an exemption from the Minimum Wage Act to reduce wages by 30%. Although clients still rely on disability pensions in addition to their wages, the director acknowledges that the center currently pays them only very low wages.

Fig. 4.15 Pen for goats

The workshop is situated next to a group home and a daycare facility, each accommodating 40 residents, mostly with mental, intellectual, or developmental disabilities. Initially, many residents participated in farmwork but found it too laborious, particularly due to the summer heat. Furthermore, the wages seemed disproportionately low for the work. As a result, only one client continues to work on the farm. This client now applies labels to cheese products under the guidance of the aforementioned staff member. Out of 40 residents and 40 regular visitors, only 18 are engaged in farm-related tasks, such as preparing meals for the group home, processing farm vegetables, and cleaning eggs. Few clients now participate in outdoor farmwork, while others spend their time resting without assigned tasks.

The vegetable team performs various tasks, including manuring, weeding, and packaging. Clients are capable of completing these tasks, except for those requiring judgment. Without guidance, they would struggle to determine which items to package or reject. On average, clients work at 40% of the efficiency of staff members due to lower dexterity. Harvesting tasks are exclusively assigned to staff members, as clients arrive late in the morning, when it is too hot for harvesting, and because the task requires judgment regarding the timing of harvest. Standardized tasks such as labeling, packaging, and egg polishing are entrusted to clients. This task allocation system developed quickly, as staff members were able to identify tasks that were difficult or easy for clients after a few trials. A few clients developed a good sense of judgment regarding accepted and substandard farm produce, while one was excluded from transplanting tasks because of difficulty in learning the correct technique. When a task proves to be too difficult for a client, staff members seek to assign them a more suitable task instead of attempting to teach them the original task. As the director explained, the inability to perform a task is not due to a lack of motivation or curiosity but arises from inherent characteristics such as limited dexterity or physical strength. The center manager added that because many clients rarely express emotions, it is difficult to determine how they feel about their work, such as whether they experience pain or pleasure.

The center serves meals to group home residents based on orders, often using vegetables produced on the farm (Fig. 4.16). However, clients seem unaware that they are the ones producing these food materials. Leftover substandard products are occasionally distributed to commuting clients, who enjoy sharing the vegetables with their families. Nevertheless, conversations within the center rarely touches on the taste of these vegetables. Occasionally, staff members receive customer feedback about the farm's produce, but this feedback is seldom shared with the clients.

One client exhibited signs of "autonomous skill acquisition" after repeatedly observing a staff member remove axillary buds from tomato plants. The client eagerly attempted the task during the staff member's absence on long-term leave. However, the client mistakenly cut off branches that should have been left to grow, despite staff members advising that the task requires a high degree of skill. While the client's enthusiasm was praiseworthy, this incident demonstrated that the client had not yet developed the required proficiency. In principle, staff should assign tasks based on a client's strengths, though this is not always feasible. Staff members

Fig. 4.16 Refrigerated vegetables produced at the farm and reserved for meals of group home residents

often find themselves explaining the reason they assign tasks to their clients in a certain manner.

When I inquired whether staff members would try to help clients improve at tasks that they were not initially good at, the director emphasized the question's inherent misunderstanding and responded as follows:

> Indeed, we could actively work to strengthen each client's abilities, but we do not, nor do we intend to. Since these clients are also employees, our primary goal is to have them perform necessary work from a management perspective. While we want clients to improve their skills, we do not prioritize increasing the productivity of disabled individuals. Rather, we focus on creating a pleasant work environment by fostering positive relationships and engaging work tasks, while only asking them to do what they are currently capable of.

Similar to other practitioners of employment support for disabled individuals, the director did not expect clients to independently acquire skills from the outset. However, he views farmwork as distinct from other occupations. When the government mandated in 1979 that disabled children must attend special schools, there was significant debate about how disabled individuals would secure housing after graduation, whether from special or regular schools. The director considers agriculture an advantageous occupation for the disabled due to the variety of tasks involved and the opportunity for contact with nature, making the work sometimes enjoyable. In contrast, other workplaces in urban settings often offer monotonous work that even non-disabled individuals would find difficult to endure for 8 h a day, 5 days a week.

The director believes the "partnership between agriculture and welfare" is valuable, as everyone requires food and supportive relationships when they are weakening. Ideally, the way people live cooperatively should also benefit others. Producing food materials requires time and care, which allows individuals facing challenges in modern society to realize their potential. With government funding to support this partnership, the director and manager aim to use this opportunity to strengthen the farm's management system. While the slogan may seem dubious, as the government appears to promote it for its own benefit, the director points out that the agricultural sector is struggling to sustain even large-scale industrial operations. Meanwhile, the social welfare budget is ballooning, depleting resources. Consequently, service providers for continuous employment support face financial instability unless they generate sufficient profit to meet labor costs for clients. Nevertheless, the director remains committed to utilizing the policy to realize their goals.

Finally, they emphasized the unique significance of their activities. Villages surrounding the farm and centers are gradually emptying, creating space for outsiders who are welcomed in the area. The leadership hopes that their model will demonstrate how farmers in mountainous regions can sustain themselves through farming.

4.4 Discussion

4.4.1 Outline of Results

Case A1 share similarities with the kitchen gardens in Miyakojima, and the settlements formed by Okinawan immigrants, aside from its legal status, as mentioned in the introduction. Through trial and error, the individual in this case became self-sufficient in terms of food. His gardening activities, therefore, provide a model for autonomous skill acquisition. Additionally, he reinvested his earnings into the construction and expansion of his house. However, the municipal job creation scheme, which he relied on for income, failed to provide consistent day labor opportunities. Given this underemployment, his gardening activities can be viewed as an alternative to the conventional job creation initiatives. Unfortunately, the interview did not clarify how he developed his observational skills. He regarded them as innate, and when asked how to enhance participants' observational skills in the farm project, he suggested selecting laborers with existing observational abilities. This response was unhelpful, as the aim of the action research project was to improve farming skills, not to screen laborers.

Case A2 further exemplifies autonomous skill acquisition in two ways: the practitioner conducts firsthand observations of his garden and sells his produce to generate funds for purchasing garden materials. This case documents the process through which the individual began displaying autonomy in acquiring farming skills. However, it also indicates that most growers cannot fully follow the guidance provided by their crops. Commuting to the farm only once a week, he struggled to meet the needs of his crops, such as watering the cherry tomato plants regularly, which led to them wilting in his first year. Over time, he learned to address such issues without needing more frequent visits. Had he attended the farm more often, he could have produced more and realized higher sales by better utilizing his skills.

Cases B1 and B2 involve farms that host individuals in need, such as the homeless and long-term unemployed. In Case B1, the director did not specifically aim to improve the farming skills of the participants but tried to distribute farm sales among them to use as revolving funds for future farming activities. In Case B2, the training focused on preparing participants for salaried agricultural work, rather than fostering independence as gardeners. The director financed the training through other business ventures. Outside the formal welfare system for people with disabilities, these practitioners would have needed to motivate or compensate themselves by finding ways to raise funds. However, they were resistant to the idea of increasing farm productivity by fostering greater autonomy among the participants.

The remaining cases operated within the framework of employment support for people with disabilities. In Case B3, a client displayed a level of autonomy beyond what the director had anticipated, except in the area of selling produce. The client's rare combination of autonomy and disability created an atypical dynamic with staff members regarding her garden work. In Case B4, the practitioners assumed a lack of autonomy among their clients and addressed this by leveraging various systems to encourage self-reliance, in partnership with other organizations and local senior volunteers.

Cases B5 and B6 involved farms on land inherited by the directors from previous generations. Unlike the situations in Cases B3 and B4, where idle farmland was utilized, these directors maintained their land before considering job creation. Both directors formed connections with disabled individuals in distinct ways but found the collaboration rewarding, despite the management challenges.

Finally, Cases B7 and B8 are noteworthy for the enthusiasm of the interviewees regarding the genuine goals of the agriculture-social welfare partnership. While Case B7 is located in the Kanto Plain, Japan's largest plain, Case B8 is situated in a mountainous area in the Kansai region. These geographical differences account for some of the variations in their perspectives. Nevertheless, the directors of both cases emphasized the importance of addressing marginalized attitudes toward those facing social exclusion and rethinking the concept of work.

4.4.2 Answers to the Initial Question

To answer the question posed at the beginning of this chapter, the relationship between human directors and instruction receivers did not significantly differ from that I was involved in, where I observed directors "passing instructions" from crops to participants. Despite the diversity in backgrounds, the relationship appeared consistent. More importantly, directors did not expect participants to be explicitly autonomous. This was true not only for organizations providing employment support to clients with disabilities but also for training programs targeting non-disabled trainees (Table 4.1). Some interviewees stated that clients should only perform tasks they were capable of at that moment. While some cases allowed exceptions for individuals who were independent from human directors, either occasionally or continuously (e.g., B1–4), these individuals could not explain what constituted such autonomy, and neither could their directors or colleagues. They merely suggested that these exceptional gardeners had a broad field of vision on the farm. This also applies to Cases A1 and A2. In this regard, the interview results offer no conclusion beyond the well-known proverb, "What one likes, one will do well."

Table 4.1 Outlines of the cases surveyed

Case No.	Corporate status or legal bases	Role of the director	What the director expects participants[a]	Extent of participants' autonomy[a]
A1	Individual	(Self-direction)	–	c
A2	Individual	(Self-direction)	–	c
B1	Workers' collective	Instruction	a ~ b	a ~ b, c in the past
B2	Non-profit organization	Instruction	a ~ b	a ~ b, occasionally c
B3	ESD[b]	Offering required information	None	c except for selling products
B4	ESD	Instruction	a	a ~ b, occasionally c
B5	ESD	Instruction	a	a ~ b
B6	ESD	Instruction	a	a
B7	ESD	Instruction	a	a ~ b
B8	ESD	Instruction	a	a

[a] (a) Participants work dependent on the director's instruction; (b) Participants work not dependent on the director's instruction but on their memories of past instructions; (c) Participants "take directions from the crops"

[b] Employment support for disabled people in conformity with the Services and Support for Persons with Disabilities Act

Nevertheless, Case A2 provided a tangible clue, as previously noted, emphasizing the importance of regular access to foodstuff in enhancing autonomy in farming. The specific ideas about food materials reflect what farm produce is desired. This clue further suggests that autonomous growers' observation skills are influenced by their off-farm perceptions of food materials and related matters. The same applies to Case A1, where the practitioner frequently exhibited curiosity about plants encountered outside of his garden. These cases indicate that autonomous farming or gardening activities foster the acquisition of autonomous skills over time. These findings suggest a potential method for enhancing observation skills, which will be explored further in the next chapter.

Looking at the interview results from a different perspective, I appear to be the only one particularly focused on autonomy. Most notably in the last case, interviewees questioned why I was so concerned with autonomy. The answer is as follows: Suppose a group of people considers someone incapable of performing a task. There are two possible counterarguments: first, it is, if not unjust, premature to assume this, and second, it may not matter if the assumption is correct. The second form is common in welfare contexts for people with disabilities (e.g., Tateiwa 2016), aligning with the interview results. The first counterargument is exemplified in the phrase "they can do it" (Chambers 1997), which underpins the concept of "participatory rural development," an alternative to authoritarian approaches common in the global South. It is this last view that resonated with me, based on my personal experiences with individuals who have cerebral palsy and various homeless laborers. Drawing on these experiences prior to initiating the job creation program, I felt compelled to remain doubtful about whether anyone is truly incapable, regardless of the circumstances.

Tateiwa Shinya, a pioneer in disability studies in Japan, has already insightfully pointed out that such job creation efforts will encounter challenges at some point (Tateiwa 2005). While meaningful to some extent, these efforts may ultimately serve as a mere complement to existing systems, which direct half-hearted efforts toward more equitable distribution of income, goods, and employment opportunities. Imagine a set of job opportunities created but deemed impossible for a job-seeker to perform. In such cases, others capable of performing the task may take over, as staff members at employment support centers often do on behalf of their clients. However, this principle does not always hold (Tateiwa 2018). No one should substitute for the individual in question if the created occupation is an end in itself rather than a means to an end. To illustrate, questioning how one can live another's life is meaningless. It is taken for granted that no one should deny someone the opportunity to live their life, regardless of their circumstances. Conversely, while it may not be universally accepted, I still believe that even staff members should not deprive their clients of the opportunity to "take directions from crops," no matter how difficult it may be for them. This is the justification I must offer here.

It should also be remembered that, as discussed in the previous chapter, several experts suggested that "taking directions from crops" involves a sense of fascination. This applies to staff members as well, who, when observing crops on the farm, make decisions about what to instruct their clients to do. Furthermore, if staff members try to prevent clients from engaging in observation, they monopolize the opportunity to experience the changing posture of crops, which should be attributed not only to their decisions but also to the labor of the clients. If fascination contributes to well-being, as will be confirmed in the next chapter, such an experience should not be taken from anyone. Vicarious observation may undermine the potential value of the agriculture-social welfare partnership.

This debate parallels past criticisms made by the disability rights movement regarding vicarious decision-making by authorities and parents, particularly in relation to the policy of compulsory segregation of disabled children in schools, established by the central government in 1979. The movement argued that such vicariousness would threaten the self-determination of disabled individuals (Watanabe 2019). However, since it is sometimes too demanding for individuals with intellectual disabilities or young children to make and voice decisions, this criticism has led to a compromise, with conditional approval for professional or parental vicariousness. The ongoing discourse has focused on supporting self-determination in a more justifiable manner (cf. Tsuda 2012; Ono and Arihara 2022). It is evident that full social participation requires the self-determination of participants. By analogy, what should be done within the framework of the agriculture-social welfare partnership remains a topic for further discussion.

4.5 Conclusion

This chapter has aimed to contextualize the findings presented in the previous chapters by comparing my research with several domestic cases, each of which approached instruction giver-receiver relationships differently. In doing so, it has led to the identification of the central question for the next chapter: If the sense of fascination triggered by the observation of crops (or livestock) enhances well-being in the context of farming activities, then the right to observation warrants protection. Moreover, what lessons can be drawn from examples of autonomous skill acquisition to achieve this goal? The following chapter will explore the connection between agriculture and social welfare in response to these questions.

Additionally, this chapter has outlined the results of trial and error among practitioners of autonomous gardening or farming, as well as the reasons why surveyed cases offer farmwork opportunities to their clients or trainees. These opportunities hold intrinsic value: The former contributes to the field of ethnobotany by providing real-time data on contemporary knowledge acquisition processes among urban gardeners and farmers, while the latter demonstrates the multifaceted nature of farming activities practiced in the name of social welfare. These findings will be revisited in a subsequent chapter.

References[4]

Chambers R (1997) Whose reality counts? Putting the first last. ITDG Publishing, London

Hardman M, Larkham PJ (2014) Informal urban agriculture: the secret lives of guerrilla gardeners. Springer

**Ishihara M (1982) The establishment of the identity of the Okinawan people in the Japan Mainland [Okinawajindekasegiijūsha no sēkatsushi to aidentiti no kakuritsu]. Bull Depart Sociol Okinawa Kokusai Univ 10(1):63–68

James S (2016) Framing on the fringe: peri-urban agriculture, cultural diversity and sustainability in Sydney. Springer

Kazemi F, Abolhassani L, Rahmati EA, Sayyad-Amin P (2018) Strategic planning for cultivation of fruit trees and shrubs in urban landscapes using the SWOT method: a case study for the city of Mashhad, Iran. Land Use Policy 70:1–9

**Kodama K (1986) An introduction to the five main highways and post stations of Edo Japan. Tokyo Bijutsu, Tokyo

Kowalski JM, Conway TM (2018) Branching out: the inclusion of urban food trees in Canadian urban forest management plans. Urban For Urban Green 45:126142

Muramatsu K (2018) Shared gardens in Strasbourg: limited sharing spaces. In: Glatron S, Granchamp L (eds) The urban garden city: shaping the city with gardens through history. Springer, pp 307–335

Nilsen M, Barnes BM (2014) The working man's green space: allotment gardens in England, France, and Germany, 1870–1919. University of Virginia Press. Kindle Edition

[4]*: in Japanese with an English title and summary, **: in Japanese with an English title, ***: in Japanese.

*Ono T, Arihara R (2022) Social work on decision making for individuals with severe motor and intellectual disabilities. J Kanagawa Univ Hum Serv 19(1):81–94

*Sugihara T (2008) Island system for development of biological resources and management of local resources: reconstructing "social unit of indigenous diversity". J Agric Dev Stud 19(1):7–14

*Sugihara T (2010) Processes and issues for the commercialization of conventional bio-resources on the southwest islands of Japan: a case study on "shi-kwa-saa" in Oogimi Village, Okinawa Island. J Rural Life Soc Japan 139:10–19

*Sugihara T, Shinzato T (2010) Cultivation and utilization of minor subsistence plant resources on the Miyako Islands of Japan. J Agric Sci Tokyo Univ Agric 55(3):227–233

***Takebe K-i (1988) The cultural history of things and people 116: road II [Mono to Ningen no Bunkashi 116 Michi II]. Hosei University Press, Tokyo

***Tateiwa S (2005) Enjoying to the limit [Genkai made Tanoshimu]. In: Tateiwa S (ed) About Hope [Kibō ni tsuite]. Seidosha, Tokyo, pp 108–125

Tateiwa S (2016) On private property. Kyoto books, Kyoto

***Tateiwa S (2018) Bodies with involuntarily motions [Hunyui no Shintai]. Seidosha, Tokyo

*Tsuda K (2012) Study regarding support based on a client-centered approach for persons with severe intellectual disabilities: discussion of the support perspective and support process. J Kansai Univ Welfare Sci 16:17–28

**Tsunashima H (2021) Report on the symposium "Toward creation of agriculture-related jobs in Kamagasaki" (URP report series 52). Osaka City University, Urban Research Plaza

***Watanabe M (2019) Are parents who commit infanticide permitted to represent their children? The Kanagawa Prefecture Branch of Aoi-Shiba Association ["Kogoroshi" Suru Oya mo Kodomo no Ishi o Ninaeru no ka? "Aoishibanokai" Kanagawakenrengōkai no Shuchō ni chakumoku Shite]. In: Kokuni Y (ed) The movement toward inclusive education regardless of disability: a pedagogical idea against compulsory provision of children with disabilities [Shōgaiji no Kyōsēkyōikuundō: Yōgogakkōgimukahantai o Meguru Kyōikushisō]. University of Tokyo Press, Tokyo, pp 118–136

Chapter 5
What Is It to Enjoy Observing the Growth of Crops or Aesthetics?

Abstract According to a key finding from Chap. 3, the model for "taking directions from crops" reveals the multifaceted nature of "observation," which also presents the main challenges in achieving autonomy in farming. It invites a discussion of the "frame problem," drawing from cognitive science and philosophy of mind. Novice horticulturists resemble AI-equipped robots, abruptly entering the real world. Their challenge lies in the realistic management of attention, which requires mental images guided by desires. When the issue is reframed as a question of intention toward a specific end product, "observation" yields an aesthetic by-product called "I-did-it experience." A paradox arises in the standard practices of "agriculture-welfare partnerships," where the teacher-learner dynamic, often lacking rational critique, may inhibit laborers from aligning their desires with the needs of crops. Consequently, the action research project shifted its focus to offering participants the opportunity to derive pleasure from sharing their harvests with others. Once any end product is identified, it becomes the starting point for a series of logical inferences, characterized by both retrospective and forward-facing aspects of "observation," which serve as repositories for I-did-it experiences. The conclusion further suggests that "observation" depends on the horticulturists' ability to memorize plant postures as idealized representations.

Keywords Aesthetic experience · Attention distribution · Desire · Frame problem · Paragon

5.1 Introduction

As discussed in the previous chapter, there are various reasons why agriculture and social welfare should be interconnected. I have emphasized the importance of autonomy, as it involves observing crop growth and experiencing a sense of fascination, as outlined in Chap. 3. Another practitioner, however, focused on offering clients enjoyable employment opportunities, which could be interpreted as fostering a "sense of liberation," as discussed in Chap. 2. Despite these differing perspectives, both agree that enabling participants to grow crops autonomously is a challenging

H. Tsunashima, *Farming and Social Inclusion in Japan*, Urban Agriculture,
https://doi.org/10.1007/978-3-032-01598-3_5

task. The key issue lies in how to address, or whether to overcome, the difficulty of developing the necessary skills. In this chapter, I argue that this challenge warrants exploration, even if the attempt to overcome it ultimately proves impossible. The primary objective is to develop a theory that addresses this difficulty, a combination of social welfare with which may seem paradoxical at present.

The chapter is structured as follows: Sect. 5.2 will integrate findings from earlier chapters to lay the groundwork for a model that explores how to "take directions from crops." This initial model will reveal the multifaceted nature of "observation" as an integral component of autonomy. Section 5.3 will examine the difficulty novice growers face in observing what they need to, borrowing concepts such as the "frame problem" and "aesthetic experience" from cognitive science and the philosophy of mind, respectively. Section 5.4 will propose a working hypothesis to address this challenge through the continuation of the action research project. Finally, Sect. 5.5 will discuss the findings, focusing on the role of memory in the model.

5.2 Groundwork for a Model of the Way to "Take Directions from Crops"

5.2.1 A Cyclical Characteristic of Horticulture

Based on the M-GTA analysis presented in Chap. 3, the process of growing crops with full autonomy involves a cyclical repetition of three stages: "observation," "decision making," and "operation." This cyclical nature results from the dual role of "observation," which not only precedes "decision making" but also follows "operation." Observation thus assumes both a forward-facing and retrospective role (Fig. 5.1). The latter is particularly important, given the inherent qualities of life forms, which are more complex to manage than inanimate materials used in other industries. This distinction differentiates farmwork from other occupations.

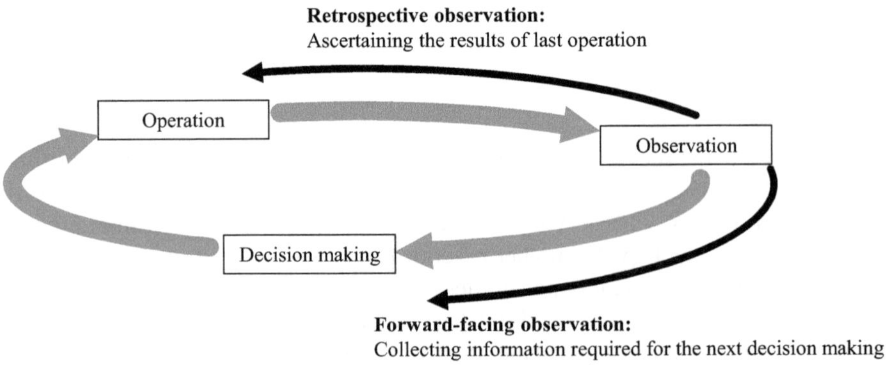

Fig. 5.1 Two-sidedness of "observation" in relation to "decision making" and "operation", which altogether comprise the cycle repeated in the act of growing crops with autonomy

Interactions between humans and non-human life forms involve uncertainty due to their individuality (Senner et al. 2015). Even a genetically homogenous group of life forms displays a range of characteristics, and individual responses to a grower's approach may vary. As a result, some, if not all, plants may deviate from the grower's intentions. Additionally, plants take time to respond to the grower's actions. For example, when sowing seeds, a grower aims to create an environment conducive to germination but must wait without guarantee that all seeds will germinate. Consequently, growers must observe whether their goal has been achieved after a certain period. Without this retrospective attitude, they could not plan future tasks in a forward-facing attitude.

The two-sided nature of observation connects the three stages into a continuous cycle. This cyclical process is essential for autonomy and reflects the ongoing relationship between growers and their crops. Consider an autonomous grower who has just completed an "operation." At this point, the current cycle is nearly complete, but the grower will begin another "observation" after some time to determine whether the previous "operation" was successful. During this time, the crop has continued to grow. If the observation indicates favorable growth, the grower can conclude that their "decisions" and "operations" were effective. If growth is less favorable, the grower will adjust their approach. In either case, the new observation will suggest further "operation," and the cycle repeats. Importantly, this repetition does not return the grower to the same place as when they started the cycle, but to a new position, marking progress along a timeline. Thus, the cycle exhibits a helical structure, with the axis aligned along time (Fig. 5.2).

The passage of time allows the grower to repeat "observation" and assess how the crops have responded to their previous actions (Fig. 5.3). This comparison between the current state of the crops (T_j) and their condition at the time of the previous observation (T_i) carries special significance. The grower's accumulated experience and knowledge about cultivation methods grow with each cycle, allowing them to improve both their crops and themselves.

This is particularly evident when the grower engages in experimental activities. As illustrated in Cases A1 and A2 in Chap. 4, growers sometimes test novel

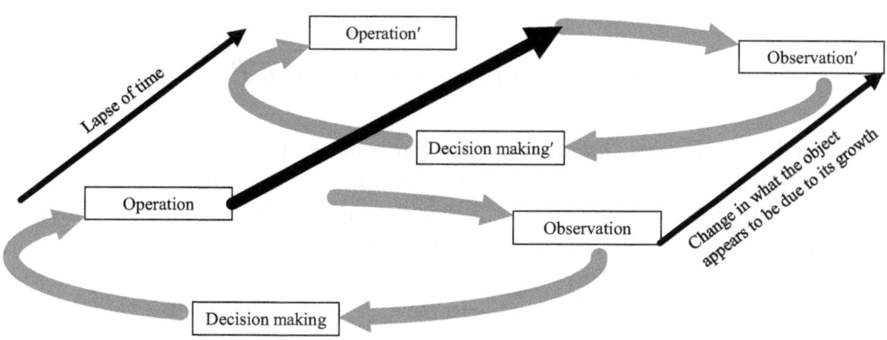

Fig. 5.2 Transition from an almost completed circle to new one occurring during a repeat of the circle comprising "observation" in relation to "decision making" and "operation"

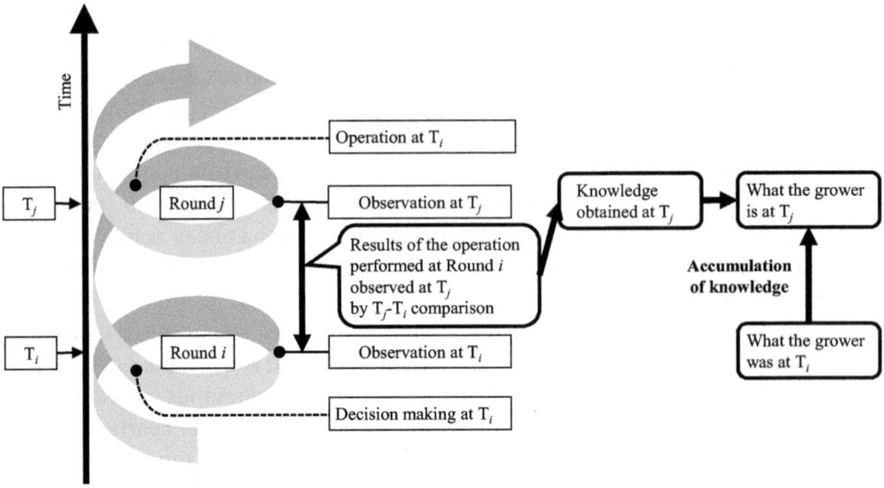

Fig. 5.3 Accumulation of knowledge as a meaning of the distance between the "observation" stage of different rounds

cultivation methods that they have developed independently or learned from other practitioners. If their subsequent observation reveals that the method was effective, they expand their repertoire of farming techniques, gaining practical knowledge. Even if the observation proves disappointing, it offers valuable insight into what should be avoided, guiding the grower through a process of trial and error. Additionally, it provides an opportunity to critically evaluate their own inventions and external knowledge. Even widely adopted farming techniques may not be suitable for every farmer, given their unique lifestyle and environment. In such cases, by discerning the core principles of a technique, growers can replace its inessential components with alternatives better suited to their circumstances. Through this process, even assimilating a small portion of "book knowledge" helps expand the grower's understanding of their crops.

5.2.2 "Observation" for Comparison Between Present and Past

Incidentally, the task of "present-past comparison," as it will be referred to hereafter, is too demanding for beginners. It requires them to accurately recall and compare the state of their crop plants at a specific point in time, relying on both their memory and previous observations. Additionally, it is difficult for growers to allocate enough time to mentally or materially capture the complete picture of their crop plants from every angle. This remains true even if they use tools like cameras with storage media to document the plants. Therefore, growers should specify the focus of their observations in advance, allowing them to compare specific aspects or parts of the plants that they deem most relevant. However, this approach is also challenging, as growers can rarely anticipate which aspects will require comparison at a future point. As I have often experienced, novice horticulturists typically realize which aspects they should have focused on only in hindsight, often triggered by an unusual occurrence. The following section will address this issue in more detail.

The current issue has been largely overlooked in past literature on horticultural observation. The three passages cited in Chap. 3 are no exception. Matsuo (2005) and Toyohara et al. (2007) assumed without question that horticultural activities always involve "interactions between perceptual and motional experiences" or that horticulturists interpret the responses of crop plants to their treatments, even though these responses can only be perceived through present-past comparison. Tsuno (1995) suggested that horticulturists compare two groups of plants, one treated and one untreated, much like a control experiment conducted by scientists. While this approach may serve the purpose of confirming whether a particular "operation" has been successfully carried out, it is disconnected from the reality faced by beginners. Control experiments, aimed at objectivity, also free observers from the reliance on memory or recording devices. However, this method is typically only feasible for advanced practitioners. At earlier stages, novice horticulturists are preoccupied with assessing the success or failure of each operation. Paradoxically, the present-past comparison of the same crop plants is often their only viable option. The turning point comes only when, for instance, they begin to question the omission or modification of certain operations, after accumulating knowledge and critically assessing what they have learned from instructors, hearsay, or media sources. Such an opportunity often leads to a deeper understanding of the necessity of each operation and facilitates knowledge assimilation.

5.2.3 "Observation" as a Comparison Between Real and Ideal

Although the present-past comparison is demanding, it alone does not allow growers to fully assess their past "operations." It only highlights whether and how the current state of the crop plants differs from their past condition, without necessarily indicating whether the change is desirable as a result of each "operation." This comparison reflects a retrospective aspect, representing one half of the two-sided nature of "observation." Thus, "observation" also involves a second comparison: between the current state of the crop plants and the expectations growers had for them after completing the relevant "operation."

The second comparison, termed the "real-ideal comparison," is more demanding for beginners, as it requires a clear understanding of the intended outcome of the current "operation." However, participants may follow instructions without fully grasping the rationale behind them, as was often the case in a previous job-creation project I managed. In hindsight, I regret not sharing the underlying principles of each "operation" more intensively with the participants. Beyond this fundamental qualification, the "real-ideal comparison" necessitates the ability to concretely visualize the ideal state of the crop plants, which growers have not yet observed in the current season. In such cases, a reference point—either from a past season or an external source—must be provided. For example, if growers can recall a successful harvest from a previous season, they may use that as a model. Alternatively, they may benefit from external sources (Fig. 5.4). The selection of such a reference requires judging what the crop plants were like at a specific past point (Y′ or Y″), based on the successful treatment they received (X′ or X″).

Cases A1 and A2 in Chap. 4 exemplify the real-ideal comparison using various reference points. These practitioners recalled the conditions of the same crop plants from previous seasons and actively sought external sources of comparison. Sometimes, at harvest, the reference may be a particular part of the crop, such as

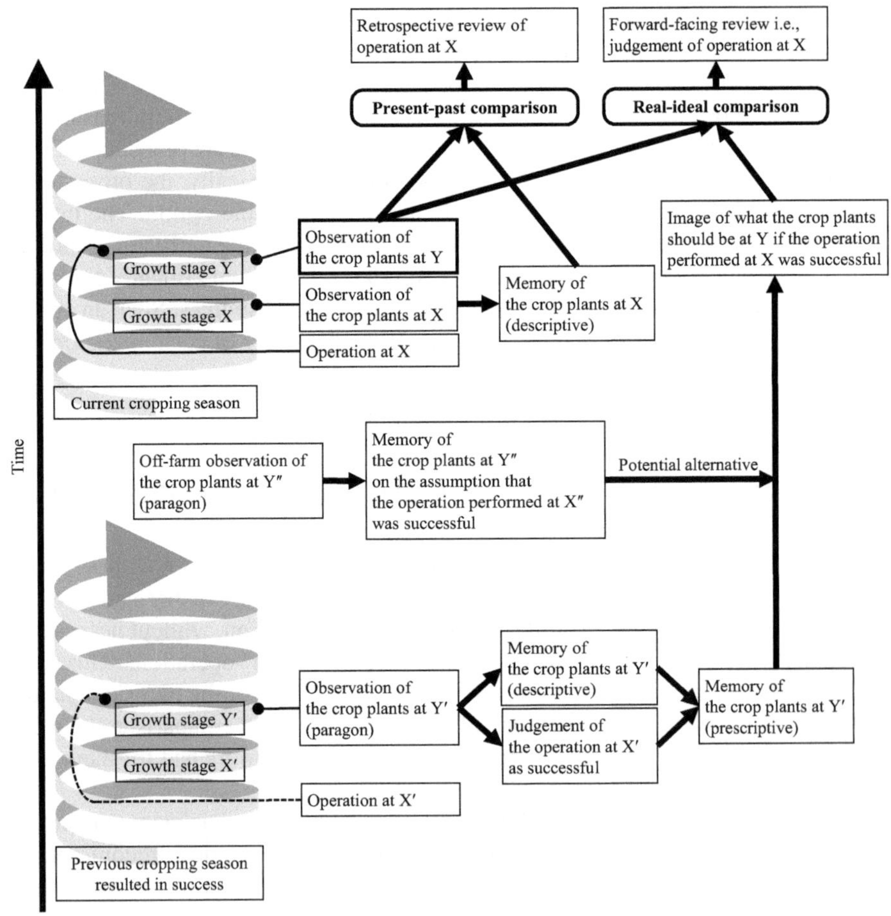

Fig. 5.4 Two-sidedness of observation comprising present-past and real-ideal comparison. X′ and Y′ are the growth stages equivalent to X and Y, respectively. Likewise, Y″ denotes the equivalent to Y, whereas X″ is an imaginary X equivalent time point

fruits displayed in stores. Other times, two-dimensional images in publications or online sources can supplement these references, though they are inherently incomplete. If growers made on-farm observations in previous seasons or referenced past off-farm observations, their recollections would form the basis for comparison during a current growing stage. The key to the real-ideal comparison lies in well-timed scrutiny, ideally completed before the current season begins, much earlier than the present-past comparison requires. Consistent with the cyclical nature of crop production, the comparison extends beyond the starting point of a single cropping season (Fig. 5.4).

The two forms of comparison also differ in terms of judgment. The present-past comparison only identifies the extent to which the crop has changed since a specific "operation" but does not assess whether the change has approached the ideal condition. This review is merely descriptive. In contrast, the real-ideal comparison requires a prescriptive approach, guided by a forward-looking perspective (Fig. 5.4). For example, suppose that after applying a mulching sheet to your vegetable plants to warm them several days ago, you observe that their stalks and leaves have grown noticeably larger. This change is striking and draws attention, reflecting the present-past comparison. However, if you were expecting flower-bud formation, this change would indicate "unproductive growth." To determine whether this is the case, you must consult an appropriate reference point to assess whether the growth aligns with your expectations. This process involves judgment, which does not automatically arise but must be actively and selectively elicited. Without such judgment, growers might lack a purposeful direction, causing the helix extension to stagnate.

Even a one-time present-past comparison can offer valuable insights to instruction receivers who have not yet fully understood the purpose of the corresponding "operation." This is particularly true when the crop plants exhibit a change so remarkable that it captures the attention of novice horticulturists. These mechanisms are exemplified by the fact that farmwork can provide clients in job-training programs with a "sense of liberation," a phenomenon also observed in some forms of green care or nature-based interventions. The next section will further explore the distinction between autonomous farming activities, which involve "judgments" derived from purposeful "operations," and passive contemplation of plant growth.

5.2.4 *From "Observation" to "Decision Making"*

The forward-facing nature of real-ideal comparisons also implies a connection to "decision making." This link between "observation" and "decision making" arises when a grower evaluates whether a previous "operation" was successful. If successful, the grower will decide to advance to the next growth stage of the crop. If unsuccessful, they must identify the cause and either apply countermeasures or repeat the same "operation."

Both approaches will lead beginners in autonomous horticultural activities into unfamiliar territory, where they will face challenges again. At this point, they may struggle to determine the next steps for their crop plants, as they no longer have a director to make decisions on their behalf. Their decisions must be based solely on their own aspirations for the plants' growth at the next stage. However, they may again find themselves unsure of how to achieve these aspirations. Without overcoming this uncertainty, knowledge and techniques will offer little help in determining what actions to take next or in following through with their plans. For beginners, it is not easy to formulate a realistic desire, given the limitations of what their crop plants can achieve. Consequently, their desire often takes on an artificial quality.

Experienced horticulturists, however, navigate this challenge with ease. They have developed clear ideas about what they can reasonably expect from their plants, rather than adhering to unrealistic desires. For example, they would never expect eggplant trees to bear gold fruit, regardless of personal wishes. These expectations are informed by observations made during previous cropping seasons or elsewhere. These models, or paragons, represent ideal outcomes at more advanced growth stages than the plants' current state. These paragons help horticulturists envision what their plants should become in the future (Fig. 5.5). As a result, these paragons, which were observed in the past (Y′), are initially used to guide decision-making in the current season (X), and later serve for comparison with the actual plants' progress at a future time (Y). The desired image of what the plants should be is shaped by these paragons. In this context, "desire" is something cultivated through learning, rather than something that naturally emerges.

Fortunately, growers in local farming communities often have access to numerous paragons, thanks to their predecessors. As a result, they are unlikely to struggle with "decision-making" unless they encounter situations that exceed their experience. These paragons may even shape the desires of growers before they begin the "decision-making" process. Just as social constructions are flexible, these desires

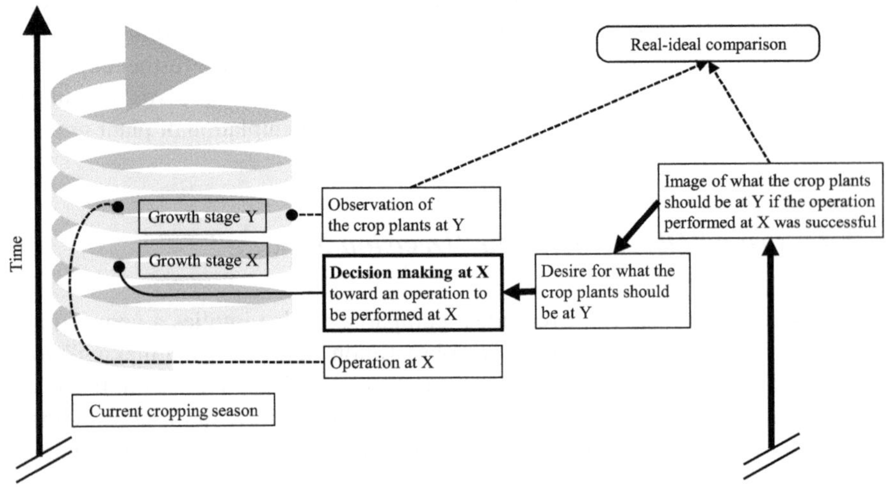

Fig. 5.5 Use of a memory of what the crop plants were in a previous cropping season for "decision making". The below half is the same as that of Fig. 5.4

are not biologically determined. For instance, most market-oriented farmers view peduncle formation in brassicaceous leafy and root vegetables as undesirable, a stance reinforced by gardening guides and instructors. However, some gardeners may actually value peduncle formation, appreciating the spikes with flower buds as an alternative food material. Thus, what growers desire from their crops at various growth stages depends on their individual purposes for cultivation.

When no relevant paragon is available at a given growth stage, growers must rely on trial and error to determine the necessary "operations." This process may lead to frustration, particularly for novice horticulturists who attempt to be autonomous without prior planning. From an observer's perspective, it may seem easier for beginners to find relevant paragons they have overlooked, which are already within their field of vision. In contrast, experienced gardeners, as illustrated in Case A1 in Chap. 4, sometimes venture into the unknown with novel plants that are unfamiliar to local predecessors. Even in such cases, they draw on their experience with closely related plants—such as varieties of the same species or species within the same genus—to make analogical inferences that guide their heuristic approaches. With accumulated successful experiences, they create new paragons that can be applied in later seasons.

Once growers have developed a mental image of their desired crop outcome, the next challenge is to determine how to materialize that image. This is purely a technical issue, leaving no room for personal discretion. Growers must rely on knowledge gained from prior experiences or advice from more experienced colleagues or former directors. The question is thus reframed: What "operation" did the grower perform at the same growth stage in previous cropping seasons to bring about their desired outcome? If the "operation" was successful, they should repeat it; if it failed, they should adjust their approach. They must also recall what they did in previous seasons to reach the current growth stage. If they cannot remember, they may need to consult the memories of their colleagues. In this sense, "taking direction from crops" means repeating past successful experiences, either personal or shared with others, since such "directions" can only be interpreted through memories of previous interactions, much like successful communication requires prior shared knowledge.

However, there are times when growers cannot take direction from crops, particularly at the beginning of a new cropping season. With no crops to observe, their decision-making must precede all other actions, though it is motivated by experiences from previous seasons. In this context, consultation with paragons becomes crucial. Additionally, two-dimensional images or hearsay may also guide the grower's initial decisions by helping to form a vision of what they can desire from the crops. In the absence of relevant information, physical materials such as seeds or seedlings of unknown species may encourage the grower to proceed with guesswork. Without anything prior, whether mental or material it is, they would be unable to resolve on starting a new crop.

5.2.5 What the Model Implies

The model developed above reveals that autonomous farming activities, in which growers "take directions from crops," are largely shaped by the multifaceted nature of "observation." Every "observation" produces an image of the crop plants at a given moment, which can then serve three distinct purposes: as a benchmark for present-past comparisons, as an ideal image guiding decision-making for future cropping seasons, and as a standard for comparing the real and ideal outcomes of the operation following the decisions made (Fig. 5.6). If the image is obtained at growth stage Y, these three events occur at Z, X′, and Y′, respectively, along the timeline axis. Notably, the two forms of comparison do not need to be explicitly mentioned in the figures, as they naturally arise when live information prompts the observer to recall corresponding images.

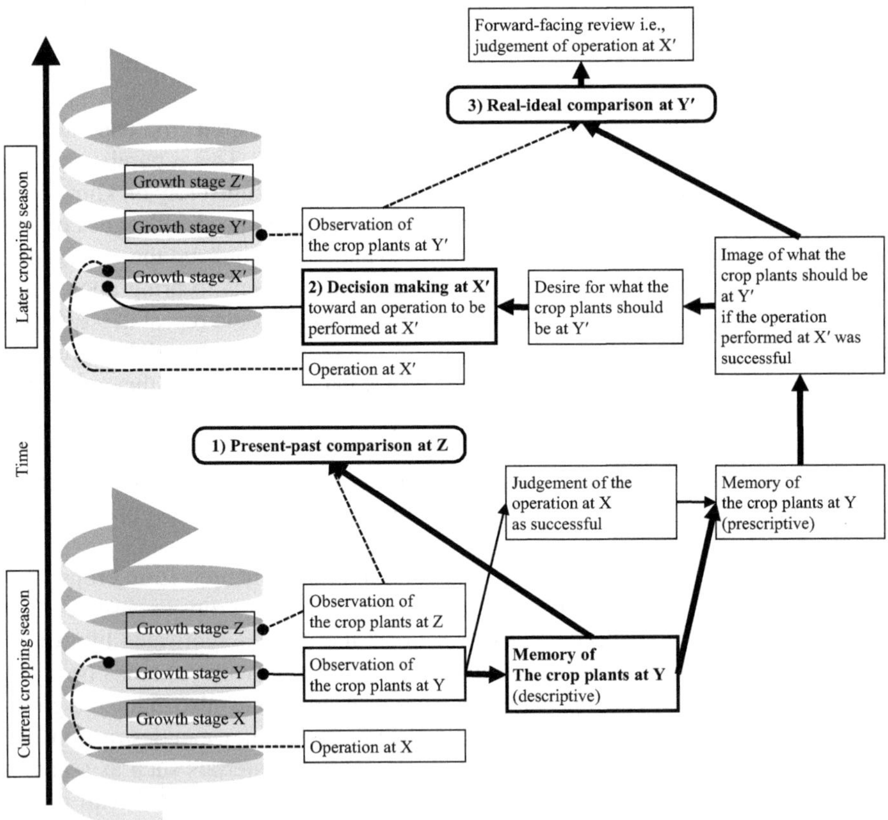

Fig. 5.6 Multiple use of the production of "observation" at a certain time point for a present-past comparison in the same cropping season on top of "decision making" and a real-ideal comparison in later cropping seasons. X′, Y′, and Z′ are the growth stages equivalent to X, Y, and Z, respectively

This analysis also highlights two key features of "observation." First, the images imprinted into the consciousness of growers at each growth stage are vital for autonomous horticultural activities in subsequent cropping seasons. As such, the preparation for the upcoming season should begin no later than the previous season. This insight is crucial for individuals training to become autonomous horticulturists. By retaining these images over several cropping seasons to come, they will eventually gain independence from their instructors. However, this earliest preparatory process is heavily reliant on perception, which is inherently challenging for trainees. Consequently, it is nearly impossible for instructors to confirm whether trainees are perceiving the intended object, especially when the percept is difficult to communicate like qualia. Moreover, it is equally impossible to force trainees to perceive something. Therefore, guiding them through this process would be a highly demanding task.

Second, while the goal of comparisons is to detect changes from the past and deviations from the ideal, neither comparison is as straightforward as "spot-the-difference puzzles". Such puzzles merely compare two static images presented side by side. Their relative ease is partly due to the lack of reliance on long-term memories or mental images. Furthermore, the solutions to these puzzles are already established. Alabi et al. (2012) argued that visual tasks are inherently challenging because they require sequential scanning of all potential features in each image. Despite their view, a thorough overhaul remains the most promising approach, though it demands sustained attention. According to Rensink et al. (1997), if observers fail to focus on a specific object, they will not notice changes to it, even if it is within their field of view. Thus, this model of comparison is not easily applicable to real-world activities like horticulture. Therefore, not only the prescriptive comparison but "observation" itself also cannot be passive perceptive activity. Connectedly, the next section will address the "frame problem."

5.3 Difficulty in Observation and Its Relation to Aesthetic Experiences

5.3.1 Revisiting the "Frame Problem"

The challenge of observing phenomena, broadly understood, can be illustrated as follows: Suppose you are observing a crop plant to gather the information necessary for performing a task within a time limit. You may direct your attention to various aspects of the plant's current state, such as stem diameter, leaf count, the size of a specific leaf, the extent of its warping, the color and texture of its surface, and the angle at which it emerges from the stem. Even for a small seedling, which consists of only a stem and a few leaves, the range of possible observations seems endless. Each of these aspects, being time-dependent, holds biological significance. However, because time is limited, it is difficult to process all of these variables, especially when the plant continues to grow in the interim. Faced with time pressure, it becomes necessary to narrow the focus to a manageable number of factors. But how can one do this effectively without a complete understanding of all aspects of the plant?

This issue, known as the "frame problem," was first addressed in the field of artificial intelligence. It soon attracted the attention of philosophers of mind and cognitive scientists who recognized that natural intelligence might face the same dilemma. As Dennett (1984) observed, this problem is exemplified in a comparison between robots and humans. What intrigues researchers is how humans seem to manage the problem unconsciously. To replicate natural intelligence in an artificial system, the system must selectively ignore most of the information it knows, focusing only on the most relevant data at any given moment (Dennett 1984). Subsequently, many scholars have explored how human cognition differs from artificial intelligence in addressing this issue (e.g., Wachsmuth 2000; Linson et al. 2018; Miracchi 2020; Roli et al. 2022). For instance, Linson et al. (2018) argued that the frame problem does not arise when an agent is continuously immersed in its environment throughout its life. In contrast, when an agent struggles to identify what external information is necessary to update its beliefs, it has effectively stopped perceiving the world. As these philosophers and scientists have suggested, this holds true for AI-equipped robots that are abruptly introduced to the virtually, if not truly, real world.

The model of autonomous farming activities outlined above reveals that the underlying issue in the difficulty of "taking directions from crops" for beginners from urban backgrounds is the frame problem. They are typically disconnected from the realities they will inevitably encounter when attempting to grow crops. Thus, it is not surprising that they experience confusion and uncertainty about what to observe and when to observe it in the field. Personally, when I was unaware of the term, most participants in the action research project—namely, laborers, trainees, and their trainers—often seemed to operate as if they were equipped with artificial intelligence, as I suggested in Chap. 3. This observation prompted me to explore theories on artificial intelligence.

In Dennett's fable of robots, the primary model focused solely on retrieving the battery to sustain its own life. Two improved successor models, in contrast, were designed to engage with their surroundings. However, both were destroyed before accomplishing their objective. Dennett appears to have used this anecdote to illustrate that any improvements would remain inherently flawed. However, a paradox remains unexplored: it was the original model, despite ultimately being destroyed, that successfully retrieved the battery from the room beyond the door. Furthermore, the original model seems more humanlike than its successors. Interestingly, no researcher in the fields of philosophy of mind or cognitive science appears to have addressed this paradox, which offers a potential clue to resolving the problem.

5.3.2 The Way Out in the Context of Autonomous Farming Activities

My personal experience from an action research project offers a useful illustration, particularly when considering two cases in reverse order of robot reform. Suppose you are growing eggplants. You arrive at the farm on a day in early summer. The first step is to observe the plants and assess any changes since your previous visit, which will guide the tasks you should address that day.

Case 1: You observe that some fruits have reached the harvesting stage. However, you are concerned about potential contingencies that may have occurred during your absence. To address this, you examine not only the fruit but also other parts of the plants, such as the leaves, stems, and roots, as well as the overall environment. As a result, you spend the entire day without completing any farm tasks. Upon reflection, you realize that you should have predetermined in advance which changes to monitor.

Case 2: You previously observed the plants beginning to bear small fruits, so you focused your farm tasks on reaching the harvesting stage. Eagerly awaiting the harvest, you concentrate mainly on the maturity of the fruit. Consequently, you overlook the possibility of other issues arising during your absence, such as pest damage or water shortage. As a result, you experience a poor harvest. Nevertheless, this outcome provides valuable lessons about the need to direct attention to factors beyond the fruit in the future, and you find yourself engaged in a trial-and-error process.

In comparison to the robot models in Dennett's apologue, your behavior in Case 1 mirrors the successor models, which, despite improvements, ultimately reached a standstill. In contrast, your behavior in Case 2 is similar to that of the original, considered primitive, model. Here, you focused narrowly on the task, resulting in some productive outcome. If the bomb in the robot apologue were less lethal, the explosion would have served as an instructive lesson, pushing the original model into a trial-and-error process, gradually expanding its focus. Although the approach in Case 1 is logically sound, the approach in Case 2 proves more effective. However, it is important to note that in Case 1, you merely dealt with the frame problem ad hoc, rather than resolved it.

The key distinction between the two cases lies in the difference in preparatory actions. Purposeful attention to a task allows for a smooth transition of focus. Since your prior farmwork was focused on the eggplant fruit, you are prepared with a clear intention to engage with the farm today. Drawing from the model developed earlier, you have an idealized image of how the eggplant fruit should appear if they have grown according to plan since the last "operation." You need not wait for further "observation"; if the fruit aligns with the ideal, you proceed with harvesting; otherwise, you delay it. This orientation toward the task is automatic, driven by various motivations, such as the desire to reap the rewards of your labor or curiosity about the crop's progress. The intention behind the previous tasks ensures continuity between past and present, as emphasized by Linson et al. (2018). In summary, intention plays a crucial role in focusing attention, sometimes more effectively than logical reasoning.

The strategy proposed for novice horticulturists to address the frame problem involves engaging them in a preparatory "operation" with a clear intention before the "observation" opportunity arises. This preemptive focus on intention helps them concentrate on the right objects during observation. Here, I designate the present attribute of "observation" as "operation-ladenness." Although this approach risks overconcentration, it can lead to a trial-and-error phase, which is essential for learning.

The model of autonomous farming activities posits that decision-making cannot occur unless individuals generate a desire, almost artificially. In this context, "desire" refers to the wish to produce a crop that matches the ideal image formed previously. A desire, in turn, requires an intention to achieve it. These two are interdependent, with desires motivating intentions, as Bagozzi (1992, 2007) argues. However, this model emphasizes that intentions, not desires, precede and shape desires. This temporal precedence of intention over desire is significant. In practice, instructors need only remind their students of the intention behind the previous "operation" to encourage the formation of appropriate desires.

Past research has not fully explored the role of desire in this process. De Sousa (1979) argued that desires are akin to emotions in that both can surpass rational decision-making in real-world situations, though he stopped short of equating desires with emotions. Ortony and Turner (1990) contended that desires are underrepresented in lists of basic emotions, not due to a lack of relevance, but because of linguistic biases. Meanwhile, Oatley and Johnson-Laird (2002) suggested that emotions are heuristics for decision-making when logical solutions are unattainable. Although they do not address the frame problem directly, their theory describes how humans resolve such problems in the real world, where solutions are often noncomputational. These philosophers agree that non-computational mechanisms help humans navigate the frame problem. Applied to autonomous farming, this mechanism corresponds to the awakening of desires stemming from a purposeful "operation" prior to the observation.

As stated in Chap. 3, "aesthetic sense" may be the apparatus for receiving "directions from crops" but a more precise definition could be provided. Serving as a partial answer to the question of how to "take directions from crops," the above discussion indicates additional factors. In summary, the key to managing the frame problem lies in the realistic management of attention during "observation." Novice horticulturists can address the frame problem by focusing their attention based on prior intentions. Therefore, the "aesthetic sense" in this context functions as representations prepared for comparison with current conditions, guided for effective attention distribution by a desire, as if a spot-the-difference puzzle is marked with highlighters.

5.3.3 *Aesthetic Experience as a By-Product of the Solution to the Frame Problem*

In general, things rarely proceed as expected, and the unexpected often occurs. As a result, a gap typically exists between what things actually are and what they are anticipated to be. This phenomenon is commonly observed among both novices and curious experts in horticulture. According to several philosophers, such a gap evokes reactive emotions (de Sousa 1979; Oatley and Johnson-Laird 2002). When a grower's expectations are met, their self-confidence increases, fostering continued trial and error and enhancing their farming skills, accompanied by excitement. However, when the outcome differs from expectations, growers often reconsider their methods. For example, a grower may feel disappointed upon finding their eggplant fruits ready for harvest but lusterless due to dehydration, signaling inadequate watering the day before. In the worst case, this may discourage them from continuing to grow eggplants. Oatley and Johnson-Laird (2002) ascribe such reactions to the working of emotions.

The distinction between the expected and the unexpected often becomes blurred when interacting with non-human life forms. These life forms inherently possess uncertainty due to their individual characteristics, as previously discussed. Once horticulturists accept unpredictability as the norm, they are less likely to expect outcomes in line with their predictions. When their pessimistic expectations are upset, the gap between prediction and reality, which can paradoxically be seen as a coincidence, becomes a welcome surprise. This surprise evokes emotion, which can inspire horticulturists to overcome their prior resignation to the challenges of growing eggplants. In that moment, the uncertainty shifts into a source of joy, potentially contributing to the well-being of the growers. In other words, the act of observation produces a by-product when the frame problem is resolved to the question of what intention the grower holds during a particular task.

I will henceforth refer to this phenomenon as "aesthetic experience" with reservations. Philosophers of art have greatly influenced debates on the meaning of this term, often focusing on the contemplation of artworks, though some also include the appreciation of landscapes (Lothian 1999; Moore 2007; Brook 2019) and others exclude it (D'Angelo 2013). Swaffield and McWilliam (2013) discuss "landscape aesthetic experience" in relation to ecosystem services, suggesting that "aesthetic experience" can refer to emotions reactive either to the formal properties of a percept or to the ideas of the perceiver reflected in the percept. Whichever attaching greater significance to, we must assume that aesthetic attributes reside only statically in the percept itself. The model of observation, however, presupposes proactive emotions in the relationship between the idea of the perceiver and the percept, which may be either a difference or coincidence depending on the perceiver's prior desires. Consequently, the earlier discussion about the appreciation of landscapes may not fully apply here. This explains why the term "aesthetic experience" must be justified as temporary. The following supplement will even clarify the unique potential of horticulture from two philosophical standpoints.

First, growing crops does not always result in the expected end products. Instead, it often leads to a grower's sympathetic attention to the plants, motivating them to engage in farm tasks with the goal of encouraging the crops to produce end products. In this way, the experience is valued intrinsically, rather than instrumentally. Many philosophers of art contend that aesthetic experience involves disinterested contemplation of an object for its own sake (Carroll 2008; Markovic 2012; Iared et al. 2016). Thus, the emotional responses that arise in horticulture can be regarded as a form of aesthetic experience. However, Carroll (2008) argues that this notion, which only identifies a necessary condition, is too limited to fully capture the essence of aesthetic experience. This critique holds for the contemplation of artworks, as the seeming disinterestedness may mask a hidden pragmatic interest. Nevertheless, when applied to autonomous farming activities, this notion proves useful because it is easy to distinguish the end products from the disinterested pleasure that emerges as a by-product. The current contrast indicates that the difference between intrinsic and instrumental value eventually becomes a matter of comparison, depending on the individual's perspective, unless an end product is premised.

Second, Prinz (2011, 2014) claims that aesthetic experiences involve a reactive emotion, specifically "wonder." This is akin to the "welcome surprise" described earlier, as both entail perceiving something unexpected in an object and appreciating some aspect of it. Prinz argues that humans cannot perceive objects without experiencing wonder, and when wonder is evoked, it draws attention to the object, further intensifying the sense of wonder. This cyclical relationship between wonder and attention does not hold, however, in the context of autonomous horticulture. If it did, there, wonder should divide into a pair of two phases: relatively earlier and later ones, based on, respectively, expectation and a reactive emotion toward the unexpected. As the earlier discussion attributed the expectation to a proactive rather than reactive emotion, which presents a contradiction with Printz's theory.

As outlined in the previous anecdotes, the properties of disinterestedness and wonder justify the use of the term "aesthetic experience" to describe the by-product of horticultural activities. Nevertheless, another important aspect of this experience arises from the "operation-ladenness" discussed earlier. The "welcome surprise" in this context is directly linked to the grower's intention when performing a prior "operation." Focused on the materialization of end products, the grower may fear failure until they notice signs of success. Upon recognizing their accomplishment, they may exclaim, "I did it!" At this point, the experience instills a sense of security, providing a reliable guide based on the grower's own practices. This experience is entirely a by-product, as it cannot occur without the initial motivation to achieve specific end products. Unlike the perception of artworks, which are crafted according to the intentions of others, this experience is grounded in the grower's personal efforts. Therefore, I propose renaming this operation-laden occurrence the "I-did-it experience."

5.3.4 Toward Systematic Formation of Longer-Lingering Desires

The formation of appropriate desires is crucial for this type of aesthetic experience and as a means to escape the frame problem. A failure in this regard can significantly hinder the autonomy of participants, a challenge that my action research project thoroughly confronted. The lack of properly cultivated desires often trapped participants within the frame problem, as they likely harbored desires unrelated to their crop plants. Since the project aimed to create alternative employment opportunities, it is unsurprising that participants sought monetary rewards for their work, as instructed by their director, a point already discussed in Chap. 3. Similarly, the trainees undergoing basic job training mentioned in Chap. 2 may have sought recognition from their trainers through allegiance. This misalignment of desires results from the uncritical relationship between instructor and learner, which leads to neglect of the farm tasks' outcomes.

For example, suppose laborers had sown seeds several days prior, following their director's instructions. The director informed them, "The current task is to germinate these seeds. Unfortunately, no one can immediately determine whether you have been successful. You will only know once you observe the germination process." The laborers seemed to understand this. However, their understanding soon proved overly theoretical. After receiving their wages, in either monetary form or another, they viewed the payment as compensation for their obedience to the director's instructions. Consequently, once the sowing of seeds was completed, the task was deemed finished, and no further engagement with it was required. The following day, whether the seeds had germinated was of no concern to the participants; they had no incentive to observe the field. Conversely, had their desire been rooted in the germination itself, they would have eagerly sought to confirm the onset of germination, much as autonomous growers do.

Although growers must adapt their desires to the various stages of crop growth, these desires form a continuous, evolving sequence. For instance, desires for germination, healthy vegetative growth, and timely flowering—all emerge from a more overarching desire for the fruit on their plants. In the context of the instruction giver-receiver relationship, sharing this continuum with trainees would likely be more effective than merely establishing discrete desires at each growth stage. Moreover, this continuum should be grounded not only in a desire for sufficient quantities and quality of the end product but also in the broader purpose of producing these products. The technical aspects of crop cultivation, governed by biological limitations, leave little room for individual discretion and therefore warrant minimal focus. In contrast, the ultimate purpose of crop production, which transcends technical constraints, remains more open-ended and often vague. Without attention to this purpose, the continuity of desires would not sustain the entire growing season, although temporary satisfaction might arise from observing healthy growth. The profound I-did-it experience emerges only when growers set a meaningful purpose, such as growing eggplants to please a specific person, and successfully achieve it.

This leads to a paradox within the standard practices of "agriculture-welfare partnerships." The inherent instruction giver-receiver relationship often inhibits workers from directing their desires toward their crops. Without such desires, however, crop cultivation cannot yield the deeply satisfying I-did-it experience. The intrinsic value of farming is thus rarely compatible with the instruction giver-receiver relationship unless it allows for critical feedback between instructors and learners. Consequently, while instructions may be a necessary evil, the challenge lies in forming an alternative. This point echoes the discussion in the final section of Chap. 3 but also introduces a new issue for research.

Agronomists have strived to translate farmers' tacit knowledge into explicit formulation, borrowing from manufacturing industries where clear verbal instructions leave no room for ambiguity. This formalization was intended to dissect real farming practices and ease technical advice from experts to farmers. This approach proved valuable for understanding farmers' decision-making patterns, especially when they seem irrational to researchers whose theoretical arguments may be impractical (Aubry et al. 1998; Dounias et al. 2002). Recently, however, the wave of formalization has extended to the agriculture-welfare partnership in response to labor shortages. To address the longstanding issue of coordinating inexperienced workers with farm tasks, the government introduced a specialized training course for coordinators in 2020 (Ministry of Agriculture, Forestry, and Fisheries 2025). In this context, "workers" refers to those who receive instructions from and work under the supervision of "practitioners," who give the instructions. The primary role of coordinators is to interpret these instructions for the workers. At times, practitioners' instructions are difficult to verbalize because they contain tacit knowledge or because senior farmers have not yet felt the need to provide detailed, specific instructions. Many practitioners and researchers report that even vague or abstract instructions can become understandable once they are broken down into manageable steps. According to a former provincial government official who completed the training course, the coordination techniques have significantly contributed to transforming more people with disabilities into valuable members of farm teams. In light of this, the present study appears somewhat regressive. While I reserve the argument that such coordination could become a form of vicariousness if it persists, as discussed in Chap. 4, I still emphasize the importance of critically examining the instruction giver-receiver relationship.

5.4 Attempt at the Formation of Appropriate Desires

5.4.1 Formation of the Conception that Growers Share the Food They Produce with Urban Dwellers

As discussed in Chap. 3, participants in the action research project sold their farm produce, engaging consumers in a tripartite relationship involving crops and their directors through direct communication at the point of sale. This interaction fosters an understanding of consumer needs and allows for the exchange of constructive feedback. Therefore, establishing continuous relationships with consumers may replace directives from human directors. In the context of promoting autonomy, participants should share their farm produce with others to receive feedback. In this scenario, selling is secondary.

The current concept draws from the "*teikei* movement", a producer-consumer partnership facilitating direct exchanges of organic produce. The movement began when urban consumers sought out agrochemical-free vegetables, initially considered substandard and unmarketable by farmers. Over time, the movement established a "face-to-face relationship" where both parties could share the produce, allowing mutual criticism based on honest feedback. This mutual trust is now seen as comparable to formal organic certification systems (cf. Akitsu and Aminaka 2010; Kim and Ikemoto 2015; Kondoh 2015). The practice has spread globally under the term "participatory guarantee system" (cf. Meinshausen and Rüegg 2010; Kaufmann et al. 2023).

Thus, extrinsic motivations, such as monetary compensation, should be replaced by something more integrated into the act of receiving feedback. One option is collectively cooking and sharing meals prepared from the harvest. This could involve serving and dining with outsiders, facilitating conversation, and offering insight into their reactions to the dishes. While selling vegetables remains a possibility, it need not be an immediate goal. Participants may first focus on acquiring the necessary knowledge and skills before considering direct sales.

Through this reasoning, the action research project, initially focused on job creation, returns to the true origin of farming activities—food production. A working hypothesis is proposed: farmwork, aimed purely at food production, may offer prospective horticulturists the opportunity to experience the intrinsic rewards of farming. To explore this further, I organized a continuation of the action research project, as detailed in the following section.

5.4.2 Methods

5.4.2.1 Formation of the Organization

A social worker in Kamagasaki, referred to as V, regularly purchased vegetables as part of the previous action research project. Since 2017, V has operated a temporary lunch café, preparing dishes from these vegetables and serving them on a voluntary basis. Located on an underutilized floor of a building in the neighborhood, the café recruited welfare recipients as business partners. V aimed to provide a sense of belonging, offering an environment free from undesirable tasks and customers. Staff received meals and small monetary compensation based on hours worked, with approximately ten guests attending each session, mostly welfare recipients or V's acquaintances.

V explained her motivation for the project: "Under current conditions, only wealthy individuals have access to healthy food and farm-related recreational opportunities, particularly organic produce." She noted that needy individuals often cannot afford to participate in such activities. Meanwhile, an increasing number of urban young people have turned to farming and rural revitalization projects, which she believed could also be replicated in urban areas. She cited examples of soup kitchens that not only serve meals to the needy but also invite them to participate in meal preparation. This approach could be extended to growing food together in the field, a concept that V found compelling.

In collaboration with V and others, I began reforming the management of the existing farm in 2019. By this time, the labor wages allocated in my research budget had been exhausted, and participants on the payroll had left, which unintentionally facilitated the reform. The lunch café was renamed "Adult Cafeteria,"[1] and we continued to serve customers once a month, as V had done previously. On the day before each session, we organized farmwork, gathering participants at a local homeless shelter and transporting them to the farm via minivan. The farmwork focused on harvesting vegetables for the next day's meals. Attendees at the farmwork session were invited to the café as guests, and volunteers were welcomed to help with meal preparation.

Participants were recruited differently than in previous phases of the project, as described in Chaps. 2 and 3. We posted flyers at the homeless shelter, inviting individuals to "join us for farmwork and cooking the harvest together." Since we could not offer monetary compensation, we were no longer constrained by previous participant limitations, resulting in a more diverse group. In addition to homeless laborers, day laborers, and welfare recipients, we welcomed non-needy citizens, such as former day laborers now working as caregivers in a social welfare organization. Some participants even drove themselves to the farm. Due to increased publicity, the café also saw a rise in customer attendance.

5.4.2.2 Data Collection and Analysis

Data collection for this study began in July 2019 and continued until May 2023 to assess whether the current initiative would achieve its intended outcomes. At the end of each farmwork session, I conducted brief interviews with each participant, asking the open-ended question, "What was new for you today?" This general query was designed to avoid leading responses. In total, I gathered 173 responses from 43 participants. Among them, 23 participants only attended once, and 6 only twice, respectively. The most frequent participant, however, was interviewed 40 times (Fig. 5.7). Additionally, a newspaper reporter and a television crew visited the farm in January and December 2020, respectively, and conducted interviews with several

[1] The term "Adult Cafeteria (*otona shokudō*)" is a parody of "children's cafeteria" (*kodomo shokudō*), a form of "soup kitchens" in Japan, which has drastically grown in number since the 2010s (Tokyo Updates 2024).

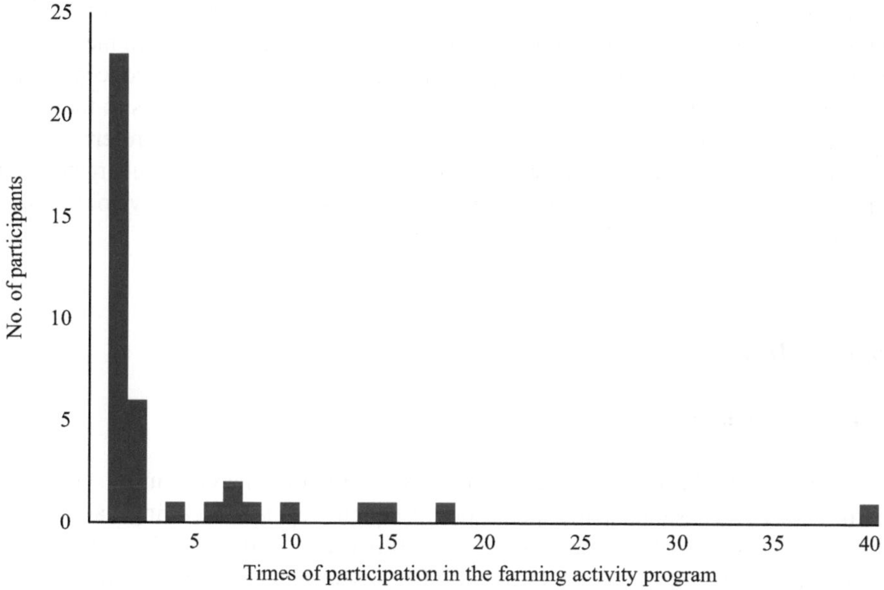

Fig. 5.7 Histogram showing the number of participants by times of participation in the farming activity program aimed at the Adult Canteen

participants. Their responses, as reported in the media, were also incorporated into the text-based data.

The primary objective of data analysis was to determine whether the model developed in previous sections accurately reflects the participants' experiences. This analysis involved exploring how participants described phenomena such as the "I-did-it experience" and its associated causative factors. I employed the coding process outlined in the M-GTA framework (see Chap. 3), bypassing the "open coding" stage and beginning directly with "selective coding." This approach was chosen because the model had already generated an adequate number of concepts that effectively represent these phenomena and define the mechanisms by which they occur. The "selective coding" stage refined these "model-generated concepts" into more specific, "text-derived concepts."

This analytical approach is similar to "thematic analysis" (Braun and Clarke 2006), which permits a "theory-driven" approach to data analysis. Furthermore, since this study involves the deliberate activation of experiences that are uncommon in the participants' daily lives, the data set is smaller than those typically used in M-GTA. As a result, the method aligns more closely with "SCAT (Steps for Coding and Theorization)." According to proponents, SCAT does not require the same level of stratification by inclusion relationships as M-GTA or thematic analysis (Otani 2008).

To initiate the "selective coding" process, I first extracted "model-generated concepts" from the model, as described later. During coding, I kept participant data separate to trace changes in their views of food and farmwork over time. This approach, though somewhat atypical in other analytic methods, allowed me to associate each experience with its corresponding participant. The terms "concepts" and "themes" are used interchangeably: in this analysis, "concepts" refers to the products of coding in the M-GTA framework, while in SCAT, they are referred to as "themes." Finally, I visualized the relationships among both "model-generated" and "text-derived" concepts in a "storyline" format. As a supplementary tool, I also utilized KH Coder 3, a text-mining software.

5.4.3 Interview Results

5.4.3.1 Participant C1

Participant C1, the most frequent interviewee, initially shared similar views on farmwork as occasional participants or infrequent repeaters. During early interviews, he described his experiences with farmwork as "absorbing," "productive," and giving him "a sense of achievement." These sentiments persisted until his sixth participation. Afterward, his motivation shifted from commercial profit to an intrinsic anticipation of the outcomes of his efforts, such as when the seeds he sowed would germinate, when the vegetables he tended would be ready for harvest, and when he could cook and eat the crops himself.

During his next visit, C1 experienced the first harvest of potatoes, which he had planted on his own. He emphasized that prior to this, he had only harvested crops planted by others. A week later, he reiterated his enjoyment of the experience, saying, "So far, what I reap had been what others planted, but today I finally harvested what I planted on my own. … I was so glad. I would be a horticulturist if I continued to do so for three years." By his 13th session, he described a sense of surprise, noting that what had attracted his attention was now connected to his previous actions.

Reflecting on his earlier experiences during his 12th session, C1 noted, "Now I do as I am told because I do not know what I should do. If I had no one to tell me, I would consult the Internet as I do for cooking." He realized that his difficulty in mastering farmwork stemmed from his passive approach—he had been following instructions without thinking critically about what he should do. He likened this to a situation where he could find no way to his destination without a guide. His comments revealed a tendency for instruction-based learning to result in a lack of independent thought during farmwork sessions.

After a year of participation, C1 established "his own garden" in a corner of the present farm, intending to continue learning and experimenting with vegetable cultivation. At this point, he had not yet reached a stage where he could consistently replicate successful crops but relied on media sources to gather information. He initially planned to try growing leafy vegetables, deemed easier, and later experiment with fruit vegetables once he gained more experience. C1 also noted that he was becoming increasingly absorbed in his personal farming efforts, which led him

Fig. 5.8 Spinaches grown in the farm (**a**), and those cooked to serve at the canteen (**b**, left end)

to reassess and refine his methods. During a winter session, participants spent most of their time collecting humus as manure material, and C1 began to understand the common farmer's saying that there is always something to do on the farm. He also "constructed" a "miniature greenhouse" to facilitate germination of crops to grow from early spring.

A memorable milestone occurred on the 24th day of his participation, one and a half years after his first visit, when he harvested spinach he had grown from the start (Fig. 5.8). He proudly donated the foodstuff for the meal served the next day, commenting, "It is exciting to imagine many customers will eat these spinaches tomorrow." On the same day, he planted seed potatoes, despite it still being slightly too cold for them to survive outside. Ten days later, he reflected, "Every time I see the weather change, there are always my crops on my mind. It has been cold for the last few days, and I have been worrying about those potatoes." Farmwork had become part of his daily life, a stark contrast to when he initially viewed it as an extraordinary task. This shift led him to dig up the seed potatoes prematurely to check for sprouts, despite understanding that doing so might damage them. Upon uncovering the potatoes, he was fascinated by how the sprouts had grown uniformly upward from the soil without any guidance.

He also found enjoyment in unexpected events, indicating his ability to compare the present with the ideal. On his 29th session, he described the germination of pumpkin as "earlier than I had anticipated." However, he faced challenges with other vegetables. For example, he had sown bitter gourd seeds as early as in March following instructions printed on the seed package, only to later remember that they should have been planted after warmer weather according to some websites. He also struggled to distinguish vegetable plants from weeds during the sprouting phase, as he lacked mental models for this stage of growth. However, once he gained these mental models, he was able to recognize seedlings, such as okra and soybean, which had seemed taking forever to germinate, expressing his delight with the words, "I thought they would never sprout anymore."

When his broad beans were having another harvesting season, he noticed that the crop was smaller than the previous year, when he had followed instructions from me. Nevertheless, he judged that he performed better at broad beans than other vegetables, seemingly because he was able to assess the current situation based on his memories of previous cropping seasons. He remembered how long it had taken for broad beans to germinate. In the following autumn, he again sowed the crop and realized that it took a week longer to germinate.

As he gained experience, he noticed signs of trouble, such as the unhealthy appearance of eggplant seedlings. This observation prompted him to pay more attention to the firmness of leaves and the angle at which they grew. A week later, he noted, "The leaves were standing erect since those plants sent roots down into the soil. They flowered furthermore." Meanwhile, his pumpkins had withered, and he attempted to infer the cause. He stated:

> I thought something must have been wrong from the beginning. Despite that, I have no idea of the reason for the failure. When I planted the crop in a planter long before, it bore big leaves and flowers. This time, I just followed the same procedure with seeds I purchased at a greengrocer. I will try again next year after sourcing seeds from a different store. Still, I also suspect that it might have been a matter of fertilization.

His real-ideal comparisons finally motivated him to learn by trial and error, inspiring him to swiftly deliberate an experimental attempt in later cropping seasons. In summary, his shift toward autonomous gardening, a precursor to the I-did-it experience, was driven by his realization that he had lacked the opportunity to translate his intentions into farm tasks. Initially, he missed the chance to harvest crops he had planted, but he longed for the sense of ownership that comes from independent decision-making. The canteen project, which allowed him to share his harvest with others, provided him with the I-did-it experience, resulting in even further encouragement. Although he initially relied on media sources for guidance, over time he began to learn from direct observations of the farm, enhancing his knowledge and skills.

5.4.3.2 Participant C2

The third-most frequent interviewee exhibited similar tendencies to Participant C1 during the early days of his participation. He remarked, "I enjoyed myself," "I learned a lot and it will be productive," and "I just like to play with soil." However, after harvesting radishes on his fifth day, he experienced a shift in perspective. In the following session, he disclosed this change, stating, "You can eat such a vegetable without cooking. This is also true for cucumbers and tomatoes, whose seedlings we just planted. I look forward to the harvest." He continued, "The farmwork is indeed physically demanding for me, but I do not dislike it. I really appreciate the value of vegetables," and added, "I would be happier if I could taste the harvests of the day in addition to tending the vegetables. Still, I will want takeouts if the harvests allow for raw consumption." Initially, he seemed to accept farmwork as a form of hospitality, rather than seeking to fulfill his own desires.

A turning point occurred when he transplanted seedlings of tomatoes, cucumbers, and other summer vegetables. His enthusiasm grew, especially regarding the anticipation of harvesting tomatoes. I recommended that he cultivate a ridge as he

Fig. 5.9 Tomato plants damaged by wild boars

wished. He planted red-leaf lettuce and successfully harvested it. Although he may have appeared overly focused on farm produce, he expressed a desire to share the harvest with other participants and cafeteria customers rather than keeping it for himself. Having already enjoyed cucumbers, he had yet to experience the harvest of tomatoes. He described his feelings at that time: "I expected the tomatoes to have fully matured, but they were not yet. They are quite delicate. I am eager for their fruits." Unfortunately, wild bears invaded the farm and damaged many tomato plants before he returned (Fig. 5.9). He lamented, "I feel sad because most of the tomatoes that we could have harvested were lost. I only got two fruits. I ate them immediately, and they were tasty."

His participation lasted less than a year due to job-hunting activities; all the above occurred in his first year. Thus, he lacked ideal benchmarks for comparisons, relying on memories of what he had eaten raw before. He also turned to two-dimensional information sources for guidance, much like Participant C1 when starting with the dedicated plot. For example, he complained, "The clay soil disappoints me when I till the field. I hope it will improve," after learning through online videos that ideal soil for upland fields is less sticky and better-drained than the farm's current soil. This indicates that fine video images serve as useful benchmarks for beginners. After tasting cucumbers, he noted, "Comparing the two, the smaller one was tastier." He could utilize this observation as a benchmark if given the opportunity to grow the crop again.

Taken together, he repeatedly gained the I-did-it experience. The first instance occurred when he harvested lettuce he had grown from the start, though he was not particularly fussy about it. The second occurred when he tasted two tomato fruits, despite the crop being a joint effort involving other participants. The difference between these occasions lay in whether a benchmark-induced mental image motivated his actions. The latter experience spurred him to develop a specific intention-desire complex in performing farm tasks. This suggests that some benchmarks are easily accessible, especially when the act of consumption is involved, as demonstrated in Case A2 in Chap. 4. In other words, such mental images may precede the formation of intention-desire complexes in novice horticulturists, and the canteen project transformed participants' memories of ready-to-eat crops into actionable benchmarks.

5.4.3.3 Participant C3

The second-most frequent interviewee had been cultivating a small garden near Kamagasaki to "prepare for the coming food crisis," as he put it. He added, "I do not want to scramble for food. Therefore, we should learn how to grow." To enhance his gardening skills, he participated in the farm and the canteen. Aside from statements in interviews, he frequently asked about the necessity of specific tasks while working alongside others.

He was curious about how crop growth differed on the present farm compared to his own garden. Early in his participation, he made the following observations:

> Potato seed tubers are buried more shallowly here than I usually do. I want to know how the size of the new potatoes will vary depending on the depth.
>
> I didn't know taro had such large leaves… I've never seen such gapless mulching with nonwoven fabric like this. I thought you had to install tunnels to mulch Chinese cabbage and lettuce in winter.
>
> I have never used bundles of rice straw to attract pea vines as done on this farm (Fig. 5.10). This method resembles *komomaki*, the practice of wrapping tree trunks in straw mats [author's note: a traditional technique once believed to trap pest insects for garden trees in winter, but recently proven ineffective]. I fear that using rice straw to attract vines may inadvertently protect pest insects instead.

The last example shows that he critically examined new techniques. He also assessed his own progress. During his fourth and sixth sessions, he commented:

> I learned that when transplanting vegetable seedlings, you should press the soil horizontally with a trowel from the side after placing the seedling in the hole. This motion creates another hole for the next seedling, making transplantation faster. When I worked as a construction laborer, I learned to master each task by observing the coworkers who completed them the fastest.
>
> Today, I found myself transplanting seedlings faster than before, although I'm still not accurate enough to avoid damaging them.

His observations also led him to refine his mental benchmarks. For instance, he noted that Welsh onion roots spread wider than he had expected. "I thought you could easily pull up Welsh onions," he remarked, suggesting that he had formed this idea based on a past experience in his childhood when his family temporarily buried whole plants for preservation. On his ninth day, he identified another useful benchmark. He said, "Eggplants here have large leaves and glossy fruits. They remind me of those in my own garden, which also produced large leaves and a successful

Fig. 5.10 Use of rice straws for pea seedlings in winter

harvest this season. I realized that you need to grow the leaves large first if you want good eggplant fruits."

Reflecting on his experience, he later said, "I now fully understand how to grow broad beans. Today's session taught me that you need provisions against the cold. It is more productive for me to watch others work closely whenever possible." He recognized the limitations of relying solely on two-dimensional information sources. The problems he encountered were of two types: those due to a lack of relevant benchmarks and those that could not be resolved even with them. His remarks reflected these challenges:

> [Unlike transplantation], direct sowing makes it difficult to distinguish crop sprouts from weeds.

> I tried to prepare broad bean seedlings for transplanting, but the sprouts rotted. I suspect the cover steamed them, or we planted the seeds upside down. I spent a lot of money on expensive seeds to compensate for my poor germination skills.

His final words in the series of interviews confirm that his journey will continue for some time: "I found that ownerless taro plants have automatically grown larger than ours nearby. There are more things that are difficult to understand as I gain experience."

In summary, he gained additional paragons, which are more reliable than hearsay, and which he would need when experimenting with unfamiliar cultivation techniques, not only knowledge of those techniques themselves. Consistent with his goal of improving his skills, his comments emphasize that paragons and techniques are inseparable. The appropriate attention given during "observation" depends on the sense of purpose maintained during the relevant "operation." He must have come to understand that "you should do or have done this to materialize something like that paragon" during the successive stages of growing certain crops. The logical connections between paragons and techniques will eventually allow growers to work backward, determining how to cultivate young seedlings based on their desired end products. Once he reaches this understanding, he will embrace the trial-and-error process more than ever, despite the challenges that may still arise.

5.4.3.4 Other Participants

Three other participants had aesthetic experiences distinct from those described earlier, who found joy in sharing food they had grown with canteen members and customers. As they cultivated crops of their own choosing, they expressed the following: "It takes a long time for spinach to germinate, which makes it even more delightful when I notice its sprouts," "It pleases me to see the plants I have been tending grow healthily. They have become surprisingly lovely, especially since my child has recently become blunt to me," "I adore watching the sprouts emerge from the seeds I sowed, especially when they develop leaves resembling those of mature plants," and "I enjoy growing living things."

These expressions not only fall under the category of "aesthetic experience" but also demonstrate an operation-laden characteristic, exemplifying the participants' I-did-it experiences. First, the more unlikely their success seemed, the greater the sense of accomplishment when it materialized. Second, these horticulturists developed an attachment to their crops, which appeared to respond to their care. Third, they sometimes felt awe at the reproductive potential of plants hidden within every seed. These reactive emotions align with the category of welcome surprise.

Other participants adopted a different perspective, positioning themselves as supporters of those striving to become autonomous horticulturists, such as Participants C1 and C2. These supporters provided objective evidence of C1's self-directed learning to overcome gaps in his memory and C2's aspiration to taste farm harvests. Examples include:

C1: "I admire his diligent study of cultivation techniques. It was clear to me that he nurtures his crops as though they were his grandchildren. This left a strong impression on me," "It was enjoyable to hear his lecture on potato cultivation," and "I learned from him how to sow spinach. There are various methods, each developed by growers through trial and error according to local conditions. There is no single correct method. This is part of what makes farming interesting."

C2: "After participating repeatedly on the farm, we can now enjoy conversations on various topics. Today, he unexpectedly shared that he had never tasted home-made dishes because his parents neglected him in childhood," and "Today we boiled rice with green peas we harvested then and there. By doing so, we were able to share and taste our harvest even with someone who has no cookware at home. I am happy to hear that he looks forward to such opportunities again."

Table 5.1 Cooccurrence relationships between the most frequently spoken words and others in the interviews with the participants who visited the farm once or twice after sessions

	Jaccard indexes				
Frequently spoken words	*Tanoshii*	*Miru*	*Tsuchi*	*Ueru*	Words with more frequent cooccurrence
Tanoshī (delightful or enjoyable)	–	0.18	0.18	0.12	None
Miru (to look at)	0.18	–	0.00	0.09	*Hajimete* (for the first time) 0.20
Tsuchi (soil)	0.18	0.00	–	0.00	*Hatake* (farmland) 0.22
Ueru (to plant)	0.12	0.09	0.00	–	*Atsui* (hot weather) 0.43 *Jagaimo* (potato) 0.33

Initially, C1 and C2 exhibited passive attitudes toward farmwork, similar to those who visited the farm only once or twice. To support this, the current paragraph presents a statistical analysis of 44 comments. The most frequent word was the adjective "*tanoshii*" (delightful, enjoyable, or pleasant), surpassing the next most frequent terms, which included the verb "*miru*" (to look at), the noun "*tsuchi*" (soil), and the verb "*ueru*" (to plant). Jaccard indexes showed no significant co-occurrence relationships between the top term and its followers, each of which was more closely linked to other specific words (Table 5.1). The interview results indicate that participants were unclear about which components of farmwork contributed to their enjoyment. It is now evident that frequent participants became more motivated toward autonomous farming only after repeated sessions in this project.

5.4.3.5 Storyline

Excluding the above interview results, the "model-generated concepts" alone form the following storyline: The I-did-it experience is a byproduct of solving the frame problem which novice horticulturists face during "observation" because they are instruction takers without desires aligned with the sense of purpose in related "operations" aimed at end products such as food. Paragons are a prerequisite for meaningful connections between purposeful "operations" and "observations." Beginners struggle to acquire these guides, as they are required to recall past events when they were unaware that they should have memorized the state of their crops at that time. Figure 5.11 illustrates this storyline alongside the "text-derived concepts" described below.

The interview results provide additional insights: First, the presence of customers behind the canteen may have contributed to some extent to the participants' desire for food production, supporting the working hypothesis proposed earlier. However, food-serving events, in particular, triggered memories of food for some

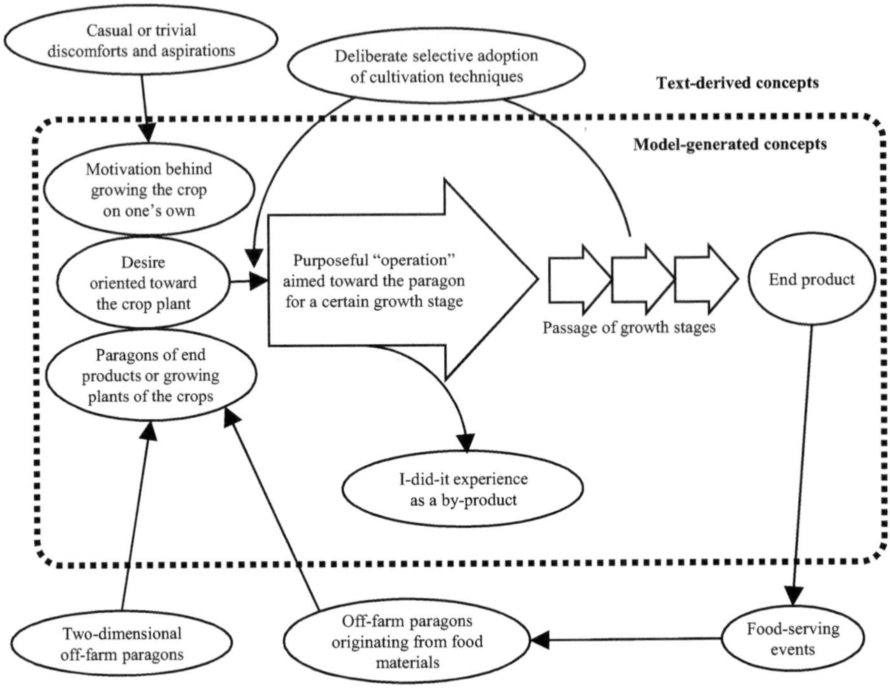

Fig. 5.11 Graphical representation of a model of the way to "take directions from crops" based on the storyline generated after the additional interview result analysis. Ellipses within and outside the bold broken line indicate model-generated and text-derived concepts, respectively

participants, reframing those memories as intention-inducing paragons. Second, casual or trivial discomforts and feelings of loss also contributed to their motivation. Third, prospective autonomous horticulturists often compensate for a lack of paragons by utilizing two-dimensional information from the media or from memories they happen to retain, which sometimes sparks a sense of purpose driven by the desire to grow crops. Over time, they replace these external paragons with those they acquire in the field. Fourth, retrospective inference connects paragons with the deliberate adoption of specific cultivation techniques. Finally, novice horticulturists undergo I-did-it experiences in various forms of operation-laden reactive emotions, each associated with a welcome surprise.

5.4.4 *Test of the Working Hypothesis*

The working hypothesis proposed at the beginning of this section was that "farmwork opportunities purely purposed for food production could offer prospective horticulturists I-did-it experiences." However, the findings from the above storyline provide additional insights. The advantage of structuring farmwork opportunities around food production lies in the clarity it provides regarding the end products for participants. Once the goal is identified—whether or not it involves food—it becomes the starting point for a series of logical inferences, accompanied by the retrospective counterpart of "observation," which serves as a repository for I-did-it experiences. In any case, the act of eating is one of the most effective ways to resolve the "frame problem" for beginners in horticulture.

5.5 Discussion and Conclusion

This chapter presents a novel model of autonomous farming activities, emphasizing two key points. First, the necessity of paragons suggests that the antecedent provides a more effective guide for producing end products than any form of instruction. However, this presents a challenge: such paragons must be prepared in advance, ideally by the preceding cropping season, or sourced from off-farm substitutes. Second, both forward-facing and retrospective forms of "observation" depend on horticulturists' ability to remember, as the comparative aspect requires mental representations of these paragons for both present-past and real-ideal comparisons. These insights will inform the themes discussed in the next chapter.

The first point aligns with the intellectual aspects of horticulture, particularly in relation to the "situated learning" movement in pedagogy. Proponents of situated learning emphasize that much of what is learned is context-specific (Kapelari 2014). For instance, Lave and Wenger (1991) in their pioneering study on apprenticeship among Vai and Gola tailors illustrate a similar concept, though they do not use the term "paragon":

> Learning processes do not merely reproduce the sequence of production processes. In fact, production steps are reversed, as apprentices begin by learning the finishing stages of producing a garment, go on to learn to sew it, and only later learn to cut it out. This pattern regularly subdivides [the learning of] each new type of garment. Reversing production steps has the effect of focusing the apprentices' attention first on the broad outlines of garment construction … Next, sewing turns their attention to the logic (order, orientation) by which different pieces are sewn together, which in turn explains why they are cut out as they are. Each step offers the unstated opportunity to consider how the previous step contributes to the present one (Lave and Wenger 1991: 72).

> Production activity-segments must be learned in different sequences than those in which a production process commonly unfolds, if peripheral, less intense, less complex, less vital tasks are learned before more central aspects of practice. … Consider, for instance, the tailors' apprentices, whose involvement starts with … finishing details on completed garments. The apprentices progressively move backward through the production process to cutting jobs (This kind of progression is quite common across cultures and historical periods) (Lave and Wenger 1991: 96).

The notions of "reversed" and "backward" learning imply that apprentices first understand the desired outcome, or the "finished" garment, and work backward to comprehend the necessary steps. This mirrors the concept of "paragons" in the model of autonomous farming activities. Beginners, when learning new tasks, rarely recognize the significance of a potential paragon until they do only with hindsight after they have completed the task. If they could recognize it beforehand, that should be attributed to their past paragon-full surroundings, such as the place they lived in or the community they belonged to. This line of reasoning underscores the value of having rural roots, as professional farmers often collect paragons unintentionally over time. The process of acquiring paragons, therefore, is inherently situated. Although Lave and Wenger developed the theory of "communities of practice," they did not acknowledge that such a learning process depends on memory. This leads to the second point of discussion.

The memory-dependent nature of "observation" implies that horticultural activities can serve as effective cognitive exercises. This idea aligns with existing literature suggesting that training can improve cognitive function (e.g., Kelly et al. 2014; Luis-Ruiz et al. 2020; Chan et al. 2024). As a result, many horticultural therapy and social/care farming practices have been developed with the assumption that such activities enhance cognitive abilities (e.g., Tu and Chiu 2020; Wang et al. 2023). For example, Kojima and Kunimi (2013) found that horticultural activities stimulate working memory systems, helping to maintain and even improve cognitive abilities in elderly individuals. Wu et al. (2020) added that, for such programs to be effective, they must be contextualized within the life experiences of participants, particularly those with rural or agricultural backgrounds. These individuals, they argue, can easily recall past experiences, which motivates them to engage in new planting activities, aligning with the principles of reminiscence therapy.

The two studies discussed here share a common premise. De Bruin et al. (2010) argue that green care farm activities demand a broader range of physical and cognitive skills and greater physical effort than typical day care services. They note that most clients have farm backgrounds, whether through upbringing, previous farming experience, or rural living, and that green care farms offer opportunities for them to relive memories from their past. Marshall (2020) similarly highlights farm-based care as a promising alternative to standard day care services for older people with dementia in rural areas. This is due to the potential of farm-based activities, characterized by "being-in-nature" and "work-kind" properties, to evoke past experiences and foster a sense of meaningful accomplishment, helping clients tangibly connect with the passage of time.

These studies suggest that horticultural activities benefit individuals blessed with rural experiences, as they can easily relate to natural environments and the activities offered. However, it can be argued that much previous research has overlooked those without such backgrounds, particularly urban dwellers. The present volume aims to explore how urban populations can be involved in farming activities and enjoy similar benefits. This leads us to the central question of the next chapter: why it is necessary to adopt a methodology that includes vulnerable urban populations in farming activities, despite the challenges involved? This approach contrasts sharply with the abovementioned counterpart encouraged by easiness, which the next chapter will address in more detail.

Wu et al. (2020) caution that decontextualized horticultural therapy may require clients to spend additional time learning unfamiliar activities, which could discourage participation. These concerns are shared by other studies on occupational therapy (e.g., Parkinson et al. 2011; Takeda et al. 2021). However, the previous section demonstrated that regular participants often contextualize their horticultural activities through practices like tasting and sharing food, as in the Adult Cafeteria project. For urban novice horticulturists, the challenge lies in accessing the paragons—essential to developing their "aesthetic sense" and observational skills. Once they learn to procure these paragons, they can begin to expand their collection and experience the sense of accomplishment that accompanies purposeful activity. This suggests that horticultural programs intended for cognitive training should consider how to link plant materials to each participant's off-farm experiences. Moreover, learning unfamiliar tasks can be a positive, rather than a detrimental, experience for participants.

This also explains why farming activities can be beneficial for individuals without rural roots, such as those affected by traumatic grief (Gorman and Cacciatore 2020). The authors emphasize that "care farming is a highly situated practice," implying that each participant will generate their own narrative through interactions with farm elements, which can foster a renewed sense of purpose. These findings reveal that clients contextualize their activities based on their personal histories. Without their agency, the farm's elements would remain unstructured, and no story would emerge. The current approach to farming activities for grief care contrasts with the abovementioned use of farming environments motivated by ease. Thus, the present volume aims to contribute to the development of more nuanced initiatives in this area.

In conclusion, the model presented here suggests that successful horticulturists cultivate an aesthetic sense as they expand their collection of paragons, which enhances their experience of participating in horticultural activities. The paragon acquisition process is "highly situated," reflecting the disparities between rural and urban landscapes, with the latter typically offering fewer paragons. Our society has shaped urban environments that lack such richness, which is a concern discussed in the present chapter.

The situatedness of landscapes emphasizes their role as learning tools. Exposure to specific types of landscapes initiates the learning process, even if the learner remains passive at first. If individuals feel they are receiving clear guidance, they experience a sense of liberation, as discussed in Chap. 2. I suspect that if local communities had appropriately arranged the spatial distribution of paragons in urban areas, individuals undergoing employability training would have experienced this sense of liberation earlier, perhaps in childhood. Bell et al. (2023) argue that "without such critical perspectives, the potential for encounters with therapeutic landscapes, including significant sensations, would become a privileged commodity for those with affluence and access, and an increasingly elusive prospect for everyone else." In this context, it is difficult to specify who "those with affluence and access" might be. The next chapter will at least explore the possibility of designing landscapes differently (cf. Thompson 2011).

Furthermore, the 1948 Universal Declaration of Human Rights asserts the right to enjoy the arts (Barrett 2022), suggesting that engagement with art is an essential component of social welfare. I argue that the I-did-it experience, which farming activities facilitate, can also be considered an integral part of this right. The model of autonomous farming activities emphasizes the importance of creating paragon-rich urban landscapes if this right is to be fully realized. The next chapter will further discuss this concept.

References[2]

Akitsu M, Aminaka N (2010) The development of farmer-consumer direct relationships in Japan: focusing on the trade of organic produce. In: Tolentino LL, Landicho LD, Wun'Gaeo S, Ikegami K (eds) Asian rural sociology IV, pp 509–520

Alabi OS, Wu X, Harter JM, Phadke M, Pinto L, Petersen H, Bass S, Keifer M, Zhong S, Healey C, Taylor RM II (2012) Comparative visualization of ensembles using ensemble surface slicing. Proc SPIE 2012:8294. https://doi.org/10.1117/12.908288

Aubry C, Papy F, Capillon A (1998) Modelling decision-making processes for annual crop management. Agric Syst 56(1):45–65

Bagozzi RP (1992) The self-regulation of attitudes, intentions, and behavior. Soc Psychol Q 55(2):178–204

Bagozzi RP (2007) The legacy of the technology acceptance model and a proposal for a paradigm shift. J Assoc Inf Syst 8(4):244–254

Barrett JM (2022) Under the aspect of eternity: the human right to enjoy the arts. Columbia J Law Arts 45:371–409

Bell SL, Hickman C, Houghton F (2023) From therapeutic landscape to therapeutic 'sensescape' experiences with nature? A scoping review. Well Being Space Soc 4:100126. https://doi.org/10.1016/j.wss.2022.100126

Braun V, Clarke V (2006) Using thematic analysis in psychology. Qual Res Psychol 3(2):77–101. https://doi.org/10.1191/1478088706qp063oa

Brook I (2019) Aesthetic appreciation of landscape. In: Howard P et al (eds) The Routledge companion to landscape studies, 2nd edn. Routledge, London, pp 39–50

Carroll N (2008) Aesthetic experience, art and artists. In: Shusterman R, Tomlin A (eds) Aesthetic experience. Routledge, New York, pp 145–165

Chan ATC, Ip RTF, Tran JYS, Chan JYC, Tsoi KKF (2024) Computerized cognitive training for memory functions in mild cognitive impairment or dementia: a systematic review and meta-analysis. NPJ Digit Med 7:1. https://doi.org/10.1038/s41746-023-00987-5

D'Angelo P (2013) Aesthetics as philosophy of experience. Proc Eur Soc Aesth 5:1–18

De Bruin S, Oosting S, van der Zijpp A, Enders-Slegers M-J, Schols J (2010) The concept of green care farms for older people with dementia. Dementia 9(1):79–128

de Sousa R (1979) The rationality of emotion. Dialogue Can Philos Rev 18(1):41–63. https://doi.org/10.1017/S0012217300047880

Dennett DC (1984) Cognitive wheels: the frame problem of AI. In: Hookway C (ed) Minds, machines and evolution. Cambridge University Press, pp 129–150

Dounias I, Aubry C, Capillon A (2002) Decision-making processes for crop management on African farms: modelling from a case study of cotton crops in northern Cameroon. Agric Syst 73:233–260

[2] *: in Japanese with an English title and summary, **: in Japanese with an English title, ***: in Japanese.

Gorman R, Cacciatore J (2020) Care-farming as a catalyst for healthy and sustainable lifestyle choices in those affected by traumatic grief. NJAS Wageningen J Life Sci 92:100399. https://doi.org/10.1016/j.njas.2020.100339
Iared VG, de Oliveira HT, Payne PG (2016) The aesthetic experience of nature and hermeneutic phenomenology. J Environ Educ 47:191–201. https://doi.org/10.1080/009588964.2015.1063472
Kapelari S (2014) Garden learning: a study on European botanic gardens' collaborative learning processes. Ubiquity Press, London
Kaufmann S, Hruschka N, Vildozo L, Vogl CR (2023) Alternative food networks in Latin America—exploring PGS (Participatory Guarantee Systems) markets and their consumers: a cross-country comparison. Agric Hum Values 40:193–216
Kelly ME, Loughrey D, Lawlor BA, Robertson IH, Walsh C, Brennan S (2014) The impact of cognitive training and mental stimulation on cognitive and everyday functioning of healthy older adults: a systematic review and meta-analysis. Aging Res Rev 15:28–43
Kim KH, Ikemoto Y (2015) Role of solidarity in organic agriculture. In: Matsui N, Ikemoto Y (eds) Solidarity economy and social business, SpringerBriefs in economics. Springer, Tokyo, pp 73–80. https://doi.org/10.1007/978-4-431-55471-4_8
Kojima H, Kunimi M (2013) The effect of horticultural activity on the cognitive performance of healthy elderly individuals. J Ther Hortic 23(2):5–18
Kondoh K (2015) The alternative food movement in Japan: challenges, limits, and resilience of the teikei system. Agric Hum Values 32:143–153
Lave J, Wenger E (1991) Situated learning: legitimate peripheral participation. Cambridge University Press
Linson A, Clark A, Ramamoorthy S, Friston K (2018) The active inference approach to ecological perception: general information dynamics for natural and artificial embodied cognition. Front Robot AI 5:21. https://doi.org/10.3389/frobt.2018.00021
Lothian A (1999) Landscape and the philosophy of aesthetic: is landscape quality inherent in the landscape or in the eye of the beholder? Landsc Urban Plan 44:177–198
Luis-Ruiz S, Caldú X, Sánchez-Castañeda C, Pueyo R, Garolera M, Jurado MÁ (2020) Is cognitive training an effective tool for improving cognitive function and real-life behavior in healthy children and adolescents? A systematic review. Neurosci Biobehav Rev 116:268–282
Markovic S (2012) Components of aesthetic experience: aesthetic fascination, aesthetic appraisal, and aesthetic emotion. i-Perception 3:1–17. https://doi.org/10.1068/i0450aap
Marshall F (2020) Farm-based care: providing meaningful activities in dementia care as an alternative to 'standard day care' in the UK. In: Innes A, Morgan D, Farmer J (eds) Remote and rural dementia care policy, research and practice. Bristol University Press, pp 185–212
***Matsuo Ē (2005) An exhortation to social horticultural science [ShakaiengĒgaku no Susume]. Rural Culture Association Japan, Tokyo
Meinshausen F, Rüegg E (2010) Alternative organic guarantee systems. In: Padel S (ed) The European Regulatory Framework and its implementation in influencing organic inspection and certification systems in the EU, pp 66–71
Miracchi L (2020) Updating the frame problem for AI research. J Artif Intell Conscious 7(2):217–230
Moore R (2007) Natural beauty: a theory of aesthetics beyond the arts. Broadview Press, Toronto
Oatley K, Johnson-Laird PN (2002) Emotion and reasoning to consistency: the case of abductive inference. In: Moore SC, Oaksford M (eds) Emotional cognition. John Benjamins Publishing Company, Amsterdam, pp 157–182
Ortony A, Turner TJ (1990) What's basic about basic emotions? Psychol Rev 97(3):315–331
*Otani T (2008) "SCAT" a qualitative data analysis method by four-step coding: easy startable and small scale data-applicable process of theorization. Bull Grad School Educ Hum Dev Educ Sci Nagoya Univ 54(2):27–44
Parkinson S, Lowe C, Vecsey T (2011) The therapeutic benefits of horticulture in a mental health service. Br J Occup Ther 74(11):525–534. https://doi.org/10.4276/030802211X13204135680901

Prinz J (2011) Emotion and aesthetic value. In: Schellekens E, Goldie P (eds) The aesthetic mind: philosophy and psychology. Oxford University Press, pp 71–88
Prinz J (2014) Seeing with feeling. In: Currie G et al (eds) Aesthetics and the science of mind. Oxford University Press, pp 143–158
Rensink RA, Kevin O'Regan J, Clark JJ (1997) To see or not to see: the need for attention to perceive changes in scenes. Psychol Sci 8(5):368–373
Roli A, Jaeger J, Kauffman SA (2022) How organisms come to know the world: fundamental limits on artificial general intelligence. Front Ecol Evol 9:806263
Senner NR, Conklin JR, Piersma T (2015) An ontogenetic perspective on individual differences. Proc R Soc B Biol Sci 282:20151050
Swaffield RS, McWilliam WJ (2013) Landscape aesthetic experience and ecosystem services. In: Dymond JR (ed) Ecosystem services in New Zealand: conditions and trends. Manaaki Whenua Press, Lincoln, New Zealand
Takeda M, Shiraiwa K, Inoue T, Imai R, Oka K, Matsuo K, Komatsu T, Kamishima T, Ashizuka A, Nakamura M (2021) Perspectives for future research on cognitive rehabilitation. Cogn Rehabil 2(1):2–19
The Ministry of Agriculture, Forestry and Fisheries (2025) The state of affairs concerning partnership between the agricultural and social welfare sectors. https://www.maff.go.jp/j/nousin/kouryu/noufuku/attach/pdf/index-11.pdf. Accessed 5 Feb 2025
Thompson CW (2011) Linking landscape and health: the recurring theme. Landsc Urban Plan 99(3–4):187–195
Tokyo Updates (2024) More than a meal: children's cafeterias build community across Japan. https://www.tokyoupdates.metro.tokyo.lg.jp/en/post-1334/. Accessed 2 Feb 2025
***Toyohara N, Yōichi I, Kae M (2007) Agricultural and horticultural activities for well-being: a manual of how to practice effortlessly [Fukushi no Tame no NōengĒkatsudō: Muri Sezu Dekiru Jissen manuaru]. Rural Culture Association Japan, Tokyo
***Tsuno Y (1995) Small agriculture: a quest from mountainous farm villages [Chīsai Nōgyō: Sankanchinōson kara no Tankyū]. Rural Culture Association Japan, Tokyo
Tu H-M, Chiu P-Y (2020) Meta-analysis of controlled trials testing horticultural therapy for the improvement of cognitive function. Sci Rep 10:14637. https://doi.org/10.1038/s41598-020-71621-7
Wachsmuth I (2000) The concept of intelligence in AI. In: Prerational intelligence: adaptive behavior and intelligent systems without symbols and logic. Prerational intelligence: interdisciplinary perspectives on the behavior of natural and artificial systems, vol 3. Springer Netherlands, pp 43–55
Wang M, Wu J, Yan H (2023) The effect of horticultural therapy on older adults in pension institutions: a systematic review. Geriatr Nurs 51:25–32. https://doi.org/10.1016/j.gerinurse.2023.02.006
Wu Y-L, Tsai W-F, Chen M-L (2020) The effect of horticultural program with life contexts on the cognitive function of the elderly in Taiwan. Problems Educ 21st Century 78(5):856–872. https://doi.org/10.33225/pec/20.78.856

Chapter 6
In Light of Research on Social/Care Farming: What Deserves International Communication

Abstract This chapter explains why the complex characteristics of horticulture must be addressed in enhancing social welfare and warrant international communication. If the agriculture-welfare partnership is to prioritize the aesthetic value of horticulture, the relevant authorities must undergo significant transformation. This necessity arises from the widespread belief that paid work is inherently painful, which may partially explain the vilification of beneficiaries. An attempt at counterargument leads to discussions on Marx's theory of alienation and primitive accumulation, as well as Okamura's proposal for mutual aid. In this context, the concept of "employment refusal support" is introduced as an alternative to simply creating employment opportunities. Next, a literature review outlines the current achievements in research on care/social farming and therapeutic horticulture. Influenced by occupational therapy traditions, research has highlighted the "meaningfulness" of horticultural activities. A philosophical approach to gardening offers a broader perspective on care/social farming. Some social work theories fill the gap between the present volume and existing literature, suggesting they share the critical perspective of traditional social work alongside social/care farming practitioners. The concluding section revisits the concept of "meaningfulness," arguing that it only becomes evident when contrasted with "meaninglessness." The following discussion challenges the mainstream view of work and care.

Keywords Alienation from work · Bullshit job · Empowerment · Primitive accumulation · Social work

6.1 Introduction

The previous chapter discussed the recent progress in the action research project, focusing on the mentally demanding aspects of horticultural activities that confront beginners. These challenges are comparable to the frame problem, which often frustrates those with natural intelligence when navigating unfamiliar territory. Despite the struggle with the uncertainty of growing life forms, autonomous horticulturists gain a sense of accomplishment through the process of producing essentials for life.

H. Tsunashima, *Farming and Social Inclusion in Japan*, Urban Agriculture,
https://doi.org/10.1007/978-3-032-01598-3_6

This unique aspect of growing crops motivates horticulturists to improve themselves. Therefore, horticultural or gardening activities should be considered valuable measures for enhancing well-being.

The findings from previous chapters are now ready to be synthesized. The project was initially designed to create alternative job opportunities in an understaffed industrial sector. However, early in the project, I observed that farming activities also foster a sense of liberation in some participants. However, this significant experience depended on the instruction giver-receiver relationships, and only after critiquing which, research can discuss the "I-did-it experience" that requires prior formation of decisions by experiencers. Beginners from urban backgrounds, in particular, struggle with this intellectually taxing demand. I attribute this discovery to the approach taken in the previous chapters, which directly addressed the difficulty of horticulture, rather than adopting ease-driven approaches exemplified in the last section of the previous chapter.

This chapter aims to reinforce this perspective, the underlying rationale has hitherto been incomplete. Accordingly, I will demonstrate that the "difficulty-conscious approach" to the agriculture-welfare partnership merits further exploration from another perspective. This could spark broader discussions on the relationship between the two sectors concerned as well as extensive societal implications. Also, I will argue that past research on care/social farming has largely overlooked the difficulty itself, as well as its connection to well-being and social welfare. To support this argument, I will review the relevant literature, particularly in light of recent research trends. This review will demonstrate that the current approach warrants international attention. Then, the discussion will shift from self-repudiation to consistency, suggesting that growing crops can significantly improve social welfare precisely *because of*, rather than *in spite of*, its inherent difficulty, despite any prior implication.

This chapter is structured as follows: Sect. 6.2 examines the findings from previous chapters and explores the meaning of growing crops in relation to past and current literature from various academic fields beyond horticultural science. Section 6.3 reviews existing literature on care/social farming and references relevant social work theories to highlight what past research has neglected. Section 6.4 unifies the previous chapters and literature review to explore untapped possibilities for collaboration between these agriculture-based activities and progressive social work.

6.2 What It Is to Grow Crops After All

6.2.1 *Domestic Equivalents to the Agriculture-Welfare Partnership*

The use of horticulture for human welfare in Japan has roots in the concept of "horticultural therapy," which was introduced from the United States in the 1990s (Matsuo 2005). This concept significantly influenced the formation of the agriculture-welfare partnership, as evidenced by a report published when the term was coined (National Institute for Rural Engineering 2008). Initially, "horticultural therapy" was aimed at populations requiring medical care but was soon applied to a broader array of initiatives, including community development. This expansion prompted the development of a more inclusive term than "therapy." In response, Matsuo (1998) introduced the concept of "horticulture for well-being",[1] which I consider equivalent to "social and therapeutic horticulture" (Sempik et al. 2005). Matsuo argued that "horticulture for well-being" encompasses all forms of horticultural applications, whereas "horticultural therapy" should specifically refer to interventions targeting individuals with impairments, disabilities, or those requiring medical or welfare services. However, some practitioners raised concerns about the current confusion between "medical" and "welfare" services as problematic.

For example, Kanda et al. (2001) conducted a survey of welfare facilities in Akita Prefecture offering horticultural activities. The main question asked whether staff considered these activities to be horticultural therapy. Few respondents affirmed this view, with most of those agreeing being from mental hospitals. Other facilities, such as those housing the elderly or people with intellectual or physical disabilities, typically did not regard their programs as horticultural therapy, despite the similarity in appearance. The most common reason for this was that these activities stemmed from another objective than therapy. This perspective was further supported by some residents with cerebral palsy at the Fuchu Center for Therapy and Education. They disagreed with the staff who imposed strict medical regulations on daily living as if they had been inpatients, which they felt humiliating (cf. Hirono 2014).[2] As disabilities do not always require medical treatment, there should be a theoretical distinction between medicine and welfare services.

[1] Matsuo (1998) introduced the term "horticultural well-being." However, I find this term unsatisfactory for two reasons: First, the phrase suggests the existence of two distinct forms of well-being—one horticultural and one non-horticultural—which seems illogical. Second, the noun "well-being" denotes a holistic concept, as horticulture alone cannot achieve it without other contributing factors. Consequently, it would be inappropriate to use "horticultural" as an adjective to modify "well-being." Therefore, I propose the alternative term "horticulture for well-being." The preposition "for" follows the construction used in "farming for health" (e.g., Hassink and van Dijk 2006).

[2] Hayashi and Okuhira (2001) describe the incident in English as a pivotal moment in the liberation movement for disabled people, although they omit the issue discussed in the text.

Thus, the above definition of "horticultural therapy" must be reconsidered. A pragmatic criterion for differentiation from "horticulture for well-being" emerged with the institutionalization of certification systems for horticultural therapists. Only those qualified under which may claim to offer horticultural therapy. Despite this, a semi-governmental consortium promoting the agriculture-welfare partnership has recently recognized some activities as equivalent, again blurring the line between the two terms. Although the distinction between "therapy" and "well-being" remains controversial, Matsuo (2005) steadfastly argued that the essence of horticulture lies in alternating mental and behavioral experiences, which I interpret as the cyclical repetition of "observation," "decision making," and "operation."

Ikegami (2013), discussing the term "potential of agriculture for well-being" coined in 2007, emphasized that agriculture inherently possesses the potential to enhance the well-being of those actively involved in it. Such potential stems from the reproductive and creative nature of agriculture, which is realized when individuals engage with living things infield, as exemplified by "horticulture for well-being" including "horticultural therapy." Hamada (2018) also views this concept as the theoretical foundation of the agriculture-welfare partnership.

6.2.2 Distinction Between Farming and Horticulture

The previous chapters have conflated the terms "(crop) farming" and "horticulture," whose meanings initially differ. This section draws a provisional distinction, which will aid later discussions. The following premise serves as the basis: while farming focuses on producing crops with satisfactory quality and quantity, horticulture involves more artistic elements. It requires specialized skills and precision, and horticulturists may emphasize the aesthetic value of their crops during the growing process.

For example, few would categorize today's highly mechanized rice production in the Global North as horticulture. In contrast, tomato cultivation exhibits characteristics more closely related to art, as the crop requires support structures to maintain proper posture and phased harvesting to ensure uniform fruit maturity. However, even in this case, a grower would still be considered a farmer rather than a horticulturist if they prioritize commercial or nutritional value over aesthetic appeal. If the crop fails to meet market standards, onlookers might dismiss the grower's efforts as not being true farming. Thus, drawing a strict distinction between the two terms is challenging.

Nevertheless, one could define "horticulture" as the practice of pausing farmwork to appreciate the beauty of crops, while "farming" is the unrelenting focus on maximizing non-aesthetic or pragmatic values. The distinction, therefore, hinges on whether the grower's motivation is driven by the crop's growth itself, irrespective of whether the purpose is home consumption or commercial production. If ornamental crops are considered, the distinction would become unnecessarily complex. Although ornamental plants serve decorative functions, the aesthetic experiences of successful growers are distinct from the "beauty" for commercial purposes expected by consumers. This aesthetic value is something that only growers can fully experience while tending to their plants.

It is important to note that farmers have likely experienced moments of aesthetic appreciation throughout agricultural history, regardless of their intentions. They may have encountered a surprising phenomenon in their crops that warranted attention, ultimately leading to a valuable discovery about the laws of nature. In such instances, the awareness that the present experience holds special significance—separate from other unexpected occurrences—would have inevitably evoked an aesthetic reaction, whether they had intended appreciation of beauty or not. Moments such as this have contributed to the ongoing development of agricultural science and technology. In this sense, what was referred to as "art" above parallels "science" in earlier times, when researchers engaged with personally relevant issues existential for themselves, unlike today's professionals who maintain a clear separation between their personal lives and their research.

As a historical fact, farming and horticulture have not completely diverged. The hypothetical situation in which farming has entirely lost its horticultural elements would leave no room for further development, only pre-established harmony, which farmers might describe as stability or banality. Thus, horticulture represents a pursuit of welcome surprise at unexpected occurrence, emerging particularly when normality has deviated from expectations, as discussed in the previous chapter. Since the notion of surprise carries a dual meaning, it is difficult to clearly distinguish between what is expected and unexpected until all aspects of the process align with the grower's expectations. Even when beginners strive to engage in farming, they often find themselves captivated by the horticultural dimension of agriculture. This is why agriculture and horticulture are inseparable, and the "potential of agriculture for well-being" must center on "horticulture for well-being."

As noted by Murray et al. (2019a), the difference between care farming and horticultural therapy lies in the therapeutic purpose of the agricultural activities involved. Agriculture's potential extends beyond horticulture, which represents only a subset of agricultural practices. While horticulture is possible with even a single house plant in an urban personal setting, agriculture encompasses a broader range of tasks, including those related to public works such as water management and land improvement. This distinction highlights the communal nature of agriculture, the possibility of which involving non-professional outsiders will be elucidated in the one after next subsection.

6.2.3 Aesthetics and the Agriculture-Welfare Partnership

If the agriculture-welfare partnership is to incorporate the aesthetic value of horticulture, the MAFF-MHLW complex must undergo a significant transformation, as this has traditionally been outside the scope of social welfare. As noted in Chap. 1, these ministries do not appear prepared for such a shift. Some pseudo-academic discourses have speculated on the potential benefits for well-being, but these deviate from the current social welfare framework. The issue lies in whether this idea should be institutionalized, as it challenges the core principles of social welfare. Hamada (2018) offers only a basic explanation of farm-based employment or vocational training, which aligns with the conventional Welfare-to-Work principle. As Takeyama (2019) suggests, proponents of the partnership hope to unlock agriculture's unrecognized potential for well-being, yet they fail to acknowledge that the existing social welfare system lacks provisions for such an initiative. The central question is whether the social welfare sector should be responsible for fulfilling agriculture's potential to benefit everyone, and why. I am concerned that the unconsidered merging of social welfare with farmwork opportunities may provoke a negative reaction toward welfare recipients.

Specifically, a haphazard appropriation of the "horticulture for well-being" concept may have unintended consequences as follows: Consider a scenario where self-proclaimed "taxpayers" criticize welfare recipients, arguing that, since they are neither disabled nor elderly, they can work. Some partnership advocates suggest that these individuals should receive vocational training or provide social services as a means of escaping their current situation. If they engage in farmwork in the hope of securing formal employment or contributing to farmland conservation, this could be seen as a commendable act. Even the critics may find it acceptable. However, an alternative scenario exists: suppose these welfare recipients participate in farmwork but derive only aesthetic value and personal happiness without achieving any tangible outcomes that would satisfy the critics. In this case, the critics may express frustration, believing that while they work hard under difficult conditions, welfare recipients receive benefits without exerting similar effort. They may argue that the welfare system should require recipients to toil rather than joyful fruits of their labor, as poverty is seen as the result of idleness rather than exploitation within a capitalist system. Moreover, a significant portion of the target population may feel particularly sensitive to such criticism, as many once took great pains to avoid becoming what they are. This narrative is not an unfounded concern within Japanese society, nor is it confined to Japan; similar patterns are observable globally, as noted by Graber (2018):

> … there is an endless drumbeat of vilification of the poor, the unemployed, and especially those on public relief—and most of people do seem to accept the basic logic of the contemporary moralists: that society is besieged by those who want something for nothing, that the poor are largely poor because they lack the will and discipline to work, that only those who do or have worked harder than they would like to do at something they would rather not be doing, preferably under a harsh taskmaker, deserve respect and consideration from their fellow citizens. As a result, the sadomasochistic element in work … has actually become central to what validates work itself. Suffering has become a badge of economic citizenship (Graber 2018: 243).

This tendency becomes even more pronounced when the beneficiaries are perceived as deserving of criticism due to an assumed sense of inferiority. The earliest advocate of "green social work" illustrates a case of green care activities as follows:

> … outdoor activities for diverting young offenders from prison, exemplify this. They encouraged personal growth in young people, and taught them skills in relating to other people and the physical environment. These endeavours declined in popularity when the tabloid press attacked them for being 'jollies' at the taxpayers' expense and the state responded by refusing to fund alternatives to prison like these (Dominelli 2012: Location 212/5758).

My concern is justified; the majority of the workforce devalues work motivated by aesthetic experiences. It is essential to theorize preventive countermeasures to anticipate the current trend of re-excluding farm trainees from citizenship. This presents a significant challenge for the agriculture-welfare partnership, which also offers the opportunity to rethink our ideas about work. As discussed in Chap. 3, the essence of autonomous farming lies in "taking directions" from less powerful life forms, not from more powerful human beings. The first half of Chap. 5 explored the I-did-it experience in this context. The issue lies in the widespread ascetic attitude toward paid work among laborers, based on the belief that workplace directions must flow from the powerful to the powerless, mirroring the oppressive nature of the existing labor market and how individuals adapt to it. Strangely, this ascetic attitude aligns with the views of critics, who believe that paid work is inherently painful. They cannot forgive welfare recipients for evading employment while experiencing happiness. Contemporary asceticism is, in part, a response to such vilification. The primary issue lies within society's majority, which views the most vulnerable as undeserving of enjoyment. This, as discussed in Chap. 1, is the true target of social welfare.

Thus, the present discussion returns to the question raised earlier: "Why agriculture for social welfare?" Instead of the ambiguous bureaucratic response given previously, this volume offers a more thorough clarification. Additionally, several alternatives proposed by practitioners in Chap. 4 provide an interpretation consistent with the theory presented in Chap. 5. The following sections will explore how the inseparability of agriculture and horticulture relates to potential shifts in our perceptions of work from those sympathetic to slanderers.

6.2.4 Link Between Horticultural Aesthetics and Social Welfare

Case B7 promoted the motto "unification of work and life," which appeared to contradict the more recent emphasis on "work-life balance." It challenges the notion that individuals must consciously reclaim this balance. However, the problem with such unification is that it can expose workers to life-threatening risks in many contemporary industries. An extreme example is the job of regularly inspecting nuclear power plant facilities, where workers inevitably face radiation exposure. Given that health damage can profoundly affect a person's entire life, their working hours must be strictly limited. Workplace harassment poses a similar threat. True "decent" work, as defined by the International Labor Organization, should integrate seamlessly into a worker's life without harm. For instance, farm laborers can align their work with the rhythms of nature—resting during rain, working early in the morning, and taking siestas in the summer. Such a working style may seem unrealistic, particularly for those shaped by urbanism, which separates them from nature's cycles.

The director argued, "Farmwork may fatigue your body, but not your mind" (see Chap. 4). In workplaces where agriculture reaches its potential, labor is not merely a commodity to be bought and sold but a result of the activation of both body and mind, deeply tied to the laborer's personal engagement. This is because, in these settings, workers cannot avoid pausing their tasks whenever the crops express the fascinating beauty, something capitalists will consider a distraction. Workers with a sensitivity to beauty are considered less efficient as laborers. Skilled farm managers, seeking to maintain strict control over their workforce, might eliminate either the beauty of the crops or the sensitive workers. Ultimately, they may find it more practical to focus solely on the end products. This situation illustrates that capital cannot fully purchase labor power under current conditions, leading to the outsourcing of the production process to individual producers, as seen in today's mainstream food distribution systems. As long as farming remains tied to horticulture, it resists the commodification and subsequent exhaustion of labor power, preserving it within the worker's possession. In this context, employees would feel no need to divide work and leisure. Agriculture, in this way, can fulfill its potential for enhancing well-being, as explored in relation to Marx's theory of alienation (cf. Crinson and Yuill 2008; Rionx et al. 2019; Oversveen 2022).

In Case B2, the belief that "to grow vegetables is to grow yourself, and it is worthwhile under any circumstances to learn how to do so" (see Chap. 4) reflects a deeper understanding than the common interpretation that an employment assistance program, if commenced with gardening activities, will somehow help individuals find employment. Some advocates argue that such programs significantly contribute to social reintegration, as evidenced by studies discussed in the concluding section of Chap. 2. According to these advocates, trainees enhance both their skills and their resolve to seek employment. However, after completing these programs, many are offered jobs with poor labor conditions. Social workers who assist homeless individuals, as mentioned in Chap. 3, lament the current labor market, where recruiters often exploit the vulnerable social positions of these job seekers. Their employment assistance actually contradicted the original aim to help service users secure good jobs, not to endure poor ones. If farming's potential for "reintegration" likewise only lead to inhumane working environments, the oppressive nature of the mainstream labor market must be addressed in the first place. The true value of farming activities should lie in true decentness rather than such potential.

The phrase "to grow vegetables…" can be reinterpreted in this light. Growers come to perceive something in the appearance of the vegetables they have cultivated—something that represents the personal, history-dependent, and scarcely transferrable skills they have gradually accumulated. As discussed in the previous chapter, urban dwellers cannot produce anything without the requisite skills, even if they suddenly acquire a piece of farmland. These skills transform the land into a productive means, allowing people to resist the consumer deskilling trend, often referred to as "McDonaldization" (Jaffe and Gertler 2006). Acquiring these skills resistantly requires abundant samples from horticultural practices or the natural world, which urban environments often lack. The scarcity of such models, as imposed by urban landscaping authorities' neglect, leads to a deskilling process in line with broader societal trends. From this perspective, any horticultural activity that exposes plant growth to the public can be appreciated. An example of this is "guerilla gardening," or the illicit cultivation of neglected urban spaces (e.g., Hung 2017; He et al. 2024). While past research has primarily examined this practice in terms of food production and city beautification (e.g., Hardman and Larkham 2014), it also serves to demonstrate to the public how plants grow.

Once farmland is combined with farming skills, the means of production can reverse the process referred to in Marxian economics as "primitive accumulation" (Glassman 2006; Perelman 2007; Nichols 2015; Chakrabarti et al. 2017). Historically, this process involved uprooting people from their means of production, such as farmland, undermining their ability to provide for themselves and forcing them into wage labor at subsistence levels (Perelman 2000). What disturbed the social workers mentioned earlier were the aftereffects of this phenomenon, as it continues as an ongoing process of capital accumulation (Fraser 2022). In response, I propose the concept of "employment 'refusal' support" rather than the conventional "employment support." This concept suggests that once potential wage laborers become deeply engaged in the aesthetic realm of agriculture—fascinated by the beauty of crops or farm animals—capital will struggle to entice them away from the land unless it offers significantly improved working conditions. In such a scenario, job seekers can reject offers of wage labor with substandard conditions. They need only sustain themselves and wait until capital becomes so desperate for workers that it has no choice but to improve these conditions. As a result, capital is forced to compete with farm life for labor, as workers remain captivated by the land while capital must find ways to appease them. This dynamic could raise the minimum standard for the entire labor market. This mysterious scenery should be the intrinsic relationship between nature and capital as it might have been observed in the early stages of mass proletarianization before it was accelerated by the Industrial Revolution.

However, this strategy of employment refusal will not necessarily lead to improved working conditions, as strikebreaking could undermine strikes unless potential job seekers unite. This is an immensely challenging task, especially in Japan, such a longtime strike-free society with sluggish industrial unionism (Sugimoto 2015). Nevertheless, agriculture still retains some potential to encourage workers toward mutual aid. As maintaining rural farming environments is too demanding for individuals or single households, farmers have organized rural communities, which are common in societies worldwide (e.g., Suehara 2006; Ochiai 2014).

These practices raise another issue linking agriculture to social welfare: the notion that individual independence is an illusion, or that to be independent is to depend on others. Nakamura (2002) was one of the first in Japan to claim that independence is, in fact, dependency. This paradox has since become well-established, particularly in social welfare studies. Okamura Shigeo, a pioneering figure in postwar Japanese social welfare, discussed labor exchange practices, known as *yui*, as an example of mutual aid but excluded them from his research (Matsumoto 2014). Instead, Okamura developed the concept of "voluntary social welfare," which he envisioned as complementing "mandatory social welfare" in a theory of social reform (Ootou 2014; Okada 2022). This incomplete theory leaves an unresolved but important question: how can mutual aid practices be integrated into the current system of social welfare?

As I mentioned earlier, practitioners can separate their horticultural activities from farming environments; rural mutual aid practices lie outside the scope of "horticulture for well-being." However, there is potential to incorporate voluntary workers, provided financial conditions allow, as wage laborers. Autonomous horticulturists often need to hire help to maintain their farming environments. Thus, the practitioners do not necessarily prioritize autonomy; they provide participants with an experience similar to the I-did-it experience. When workers perceive that their efforts have transformed the local landscape, they may experience a sense of achievement akin to the model of autonomous horticulture, as if the local community itself were a living entity. This dynamic relates to the inherent uncertainty or changeability of the environment, as discussed in the previous chapter. Consequently, effective agriculture-welfare partnerships would address both material and human aspects of farming environments. This aligns with the aforesaid practitioners' descriptions of agriculture's welfare-oriented characteristics, such as "workers can work as part of a team" and "every worker can find their niche."

In summary, the inseparability of agriculture and horticulture gives farming activities unique attributes in creating employment opportunities. This uniqueness has been overlooked in discussions on the agriculture-welfare partnership, which has overly focused on the "creation" of jobs, addressing "why work for social welfare?" rather than "why agriculture for social welfare?" This uniqueness, however, connects to issues that earlier theorists like Marx and Okamura addressed, offering important insights into, if not resolving, the latter question.

One missing link between the theories of Graber (2018) and Fraser (2022) is the joy of caring for life forms weaker than oneself, a form of the I-did-it experience. If agriculture's potential for well-being drives the partnership, the resulting jobs will differ fundamentally from the "bullshit jobs" described by Graber (2018). These jobs, emblematic of the feudal hierarchy revival in contemporary capitalism, are often pointless, yet inexplicably well paid. In contrast, farmwork produces tangible forms of wealth, thus being genuinely productive, even if poorly compensated. Fraser (2022) suggests that carework, which is essential for social reproduction, is often undervalued or excluded from the official economy. Like carework, farmwork involves dealing with uncertainty and provides aesthetic experiences. According to the present theory, the commodification of labor is unnatural in the realm of human care, as capital recognizes commercial value of only the tedious aspects of carework, neglecting its joyful elements, which are equally essential to acquisition of reward-worthy skills of carers. This oversight is often ignored in discussions on the "crisis of care" (Fraser 2022). Nevertheless, we can sketch an alternative vision for the cooperation between agriculture and social welfare, one that goes beyond merely creating paid employment opportunities, although further details will be discussed in the concluding section.

6.3 Other Possible Explanations for the Reason "Why Agriculture for Social Welfare"

6.3.1 Well-Being and Meaningfulness

The section explores the literature on international parallels to the agriculture-welfare partnership to uncover additional reasons for "why agriculture for social welfare." As briefly discussed in the introductory chapter, Sempik et al. (2005) raise a critical question: Why are horticulture and gardening effective in promoting social inclusion, whereas other therapeutic activities could potentially have similar benefits? The key inquiry is to identify the unique advantages that horticulture and gardening offer over other activities. Sempik et al. (2005) argue that the benefits of social and therapeutic horticulture, among others, are tied to the "breadth and richness" of gardening practices. They propose a "logic model" (cf. Murray et al. 2019a) to explain how health and well-being result from participation in horticultural activities or engagement with the natural environment, which provides both a backdrop for these activities and benefits in its own right. They suggest that interaction with the environment can restore attention and satisfy spiritual needs related to purpose, meaning, and connectedness to nature.

However, I find it difficult to identify any elements specifically attributable to growing crops within the model, which might support the idea that "breadth and richness" means that merely being active outdoors and in contact with nature yields several benefits. A notable gap in the model is the lack of explicit mention of the "aesthetic value of the natural environment," despite examples from clients, such as:

> I think one of the biggest things about [a] garden is the very fact that it is alive, and it responds to whatever you do. And people can see the results of their actions fairly quickly …
>
> I really like … wondering if your seeds have germinated yet, and whether one of your plants is flowering yet. There were lots of things to see each day and people would take delight if they'd found something that had suddenly come into flower and looked really beautiful. (Sempik et al. 2005: 48)

The slight in these narrative data may reflect a presumption that therapeutic horticultural or gardening activities should be collective. Thus, it is not surprising that individual experiences were scarcely mentioned. A similar perspective was explored by Fieldhouse (2003), who investigated how communal horticulture contributes to social inclusion and its potential use in occupational therapy. He viewed communal gardening as an accessible occupation due to the affordability and widespread availability of public allotment gardens.

Fieldhouse's study emphasized the primacy of subjective experience in determining the meaningfulness of occupations. For occupational therapy, meaning is a vital phenomenon influencing clients' engagement, ongoing process, and outcomes. However, therapists cannot impose personal meaning on an occupation; rather, clients assign meaning subjectively and contextually. To address this, the theme was examined phenomenologically. The analytical framework drew heavily on the Attention Restoration Theory (Kaplan 1995) and the theory of flow (Csikszentmihalyi 1992). According to Attention Restoration Theory, the natural environment aids in recovering from mental fatigue. Fascination with plants involuntarily captures attention, which is restorative for individuals fatigued by the effort of focusing on specific objects (Kaplan and Kaplan 2011). In this context, horticulture's effectiveness is linked to evolutionary mechanisms developed to cope with the environments in which early humans evolved. The flow theory suggests that "flow" is a subjective psychological state of immersion, marked by enjoyment, self-motivation, and a sense of competence during an occupation.

The interview results indicated that the interplay between attention restoration and the flow experience facilitated participants' engagement. Later stages of attention restoration involved contemplating the natural world and self-reflection on meaningful occupation, which expressed the truest self. This process alleviated interpersonal tension within the allotment group and reduced attention fatigue. In this way, the occupation, as an enabler of social inclusion, gained additional personal meaning through sociocultural connections with others and the broader community. These findings contrast sharply with Chap. 2, where absorption resulted from a sense of liberation after receiving rational instructions aligned with natural laws. In that chapter, I identified absorption as a latent essence of farming activities linked to the nature of nature. In contrast, this study emphasizes "absorption" as a precursor to several factors contributing to social inclusion.

By the early 2000s, research from various countries had demonstrated that people-plant interactions, or "working with plants," promote human well-being across different groups (Elings 2006). In addition to the Attention Restoration Theory, the then other leading theory on the mechanisms behind horticultural therapy was the psycho-evolutionary theory, which posits that humans' positive psychological responses to plants are rooted in evolution (Ulrich and Parsons 1992)—humans therefore have an inherent tendency to pay attention to certain plant and environment combinations. Fieldhouse (2003) integrated these two theories into a prototype, while Sempik et al. (2005) only implicitly referred to the Attention Restoration Theory. Irvine and Warber (2002) explicitly adopted the same theory, as cited in the article below. Both theories suggest that plants naturally attract attention from humans regardless of their disparity in socioeconomic status. However, the findings in Chap. 3 challenge this view, revealing that the instruction giver-receiver relationship may distract participants from the growth of plants they have tended. This chapter undermines the applicability of these theories. Chapter 5 further scrutinized ways to restore attention, addressing a topic that would otherwise be unnecessary. Moreover, the mechanisms leading to the diverse beneficial effects of people-plant interactions remain unclear.

It appears that Elings (2006: 52) conceived the "sense of liberation," stating, "The garden is a safe place, a friendly setting where everyone is welcome. Plants are non-judgmental, non-threatening, and non-discriminating. ... Our high-tech world is unpredictable, but plants have a fixed cycle that we can rely on." The "unpredictable" likely refers to human capriciousness, observable in workplaces where superiors exert control over subordinates, transcending natural laws. Unfortunately, the author overlooks the uncertainty inherent in plant growth, a necessary element for the I-did-it experiences discussed in Chap. 5. The author also mentions autonomous gardeners and those undergoing instruction for educational purposes but fails to address the distinction between the two groups or how the latter may transition to the former. This gap, which I address in Chaps. 3 and 5, remains an important consideration.

Meanwhile, the concept of meaningfulness was further developed in subsequent research. Gonzalez et al. (2011) argue that therapeutic horticultural activities address existential issues faced by individuals with clinical depression. Their article also presents a supplementary bibliography of conceptual foundations for horticultural therapy, outlined as follows:

First, meaningfulness motivates patients to persist long enough to achieve therapeutic benefits (Trombly 1995). Second, the authors credit Ulrich (1983) with the concept of "aesthetic experience," which, in this context, differs from the meaning of the term in Chap. 5. Ulrich argued that contemplation of both natural and man-made scenes, including those with vegetation, evokes affective emotions in observers—even if such experiences are one-time occurrences. However, it is important to note that plant growth entails delays, requiring repeated observation. Third, Unruh (2004) suggested that gardening holds significant meaning compared to other leisure activities, such as music or art, because its seasonal nature mirrors the life cycle of growth, maturation, and death. However, the meaning of gardening varies based on individual interests, past experiences, and current circumstances. While this article implicitly includes ongoing contemplation, it is not always linked to horticultural activities, which necessarily involve the act of growing plants by either self-directed or guided participants. Fourth, Waliczek et al. (2005) argued that gardens satisfy human needs across all levels of Maslow's hierarchy, from basic physiological ones to higher-level ones for societal connection and self-actualization. I view this multifaceted fulfillment as a form of meaningfulness, despite the lack of definitive information in their citation. Lastly, according to Irvine and Warber (2002), the existing literature suggests that interaction with nature positively impacts human health in various dimensions. This "interaction" spans a continuum, from passive perception of a natural setting to active engagement in gardening.

I largely agree with Gonzalez et al. (2011), although I find certain points unclear due to logical inconsistencies about what "meaningfulness" precisely entails. The authors attribute meaningfulness to "excitement and absorption in following the growth process of plants," a phrase that, though intended to provide insight, remains ambiguous. If "excitement and absorption" are meant to reflect the satisfaction derived from personal achievement, the experiencer would be the grower, as "excitement" results from operational, hands-on observation, as discussed in Chap. 5. In

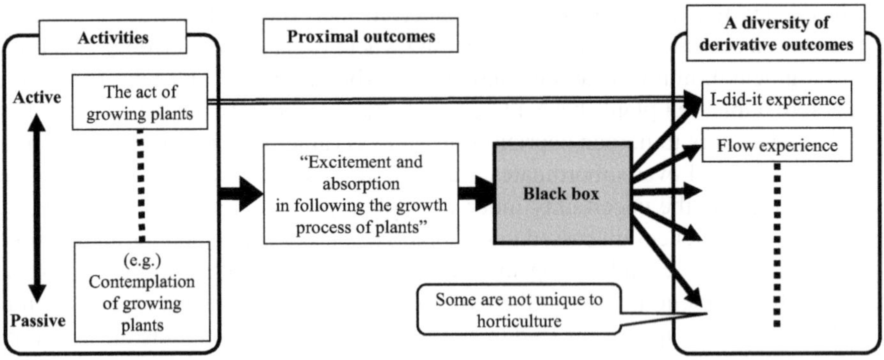

Fig. 6.1 Black box that contains the causal relationship between the multifacetedness of therapeutic horticultural activities and their diverse outcomes in Gonzalez et al. (2011)

this case, meaningfulness should be exclusively tied to the act of growing plants. However, the authors also acknowledge that therapeutic horticulture encompasses both "active" and "passive" activities. Passive activities, which may not involve direct plant care, could include simply observing growth. Therefore, "following the growth of plants" might involve passive, repeated contemplation. This phrase seems to represent something akin to the flow experience, characterized by a combination of favorable emotions, as described in Chap. 2, which could underpin a variety of therapeutic interventions. Thus, "excitement and absorption" may be basic and even commonplace outcomes, rather than those exclusively derived from the act of growing plants. This raises the fundamental question—why should growing actively be emphasized over following growth passively?

By prioritizing the "breadth and richness" of therapeutic horticultural activities, the article overlooks an explanation of how these activities generate a range of subsequent benefits. Among these is the I-did-it experience, which is grouped with other miscellaneous outcomes. As a result, a gap exists in understanding the causal relationship between the multifaceted activities and their diverse derivative outcomes, as though contained in a "black box" (Fig. 6.1).

6.3.2 *How to Evidence the Mechanisms Concerned?*

Review articles on horticultural therapy, care, and social farming began to emerge in the 2010s, as identified through a Google Scholar search. In 2000, only approximately 30 articles on horticultural therapy were available, but by 2024, this number had increased tenfold to over 600. Similarly, articles on care/social farming were scarce until 2010 but had reached 300 by 2024. This increase aligns with the growing trend toward evidence-based medicine, which emphasizes a hierarchy of clinical research designs, with systematic reviews and meta-analyses at the pinnacle (Rosner 2012). In response, research during this period aimed to provide better evidence for the effectiveness of green care interventions.

Iancu et al. (2013) conducted the first systematic review of farm-based interventions for individuals with mental disabilities. Their synthesis of quantitative and qualitative data suggests that farm-based environments facilitate social interactions, with communication with farmers and peers described as beneficial. However, it remains unclear whether these effects are specific to farm settings. The authors concluded that while high-quality studies are limited, the role of farm-based interventions in recovery and rehabilitation has only been superficially explored. Although quantitative data can help assess efficacy for certain groups, they do not explain the mechanisms by which farm-based interventions work. Qualitative data, on the other hand, attributes the therapeutic effects to the absorption of working with plants and the sense of connectedness to nature. Thus, the relationship between horticulture and therapeutic outcomes remains unclear, creating another "black box" of unexplored connections.

Similarly, Bragg and Atkins (2016) reviewed the literature on "nature-based interventions" for mental health care, including care/social farming and social and therapeutic horticulture. To streamline communication between service providers, commissioners, and clients, they propose the term "green care" to encompass nature-based activities for individuals with specific needs. They distinguish care/social farming from social and therapeutic horticulture, noting that the former involves relationships with farm community members and occurs in an informal, non-clinical context, while the latter is more structured. Both approaches, however, engage participants more actively with nature than other green care forms, which are typically limited to being in or viewing nature. The authors suggest that these hands-on interventions derive mental health and well-being benefits from the combination of the natural environment, meaningful activities, and social context, allowing service users to engage with and shape nature.

The article also references Edward Wilson's Biophilia Theory, alongside Attention Restoration and Psycho-evolutionary Theories, to support the link between horticulture and well-being, though the mechanisms remain uncertain. Thompson et al. (2012) identified three primary pathways through which the natural environment provides mental health benefits: direct restorative effects, opportunities for positive social contact, and promotion of physical activity. The authors presuppose a consensus that nature contributes to well-being, mental development, and personal fulfillment. Natural green environments offer spaces to relax, escape modern life stresses, socialize, and engage in physical activity, all of which have positive effects on mental health. The current review could thus present a revised logic model illustrating activities in the natural environment (Fig. 6.2).

Murray et al. (2019a) conducted a systematic review to develop new logic models explaining how care farming may benefit various vulnerable groups. Their model presents a layered structure that includes anticipated delivery mechanisms, intervention components, and intended outcomes. The authors mapped these mechanisms to five theoretical concepts: the restorative effects of nature, social connectedness, personal growth, physical well-being, and mental well-being. However, the relationship between many mechanisms and outcomes remains unclear (Fig. 6.3). Surprisingly, the studies reviewed did not mention the widely

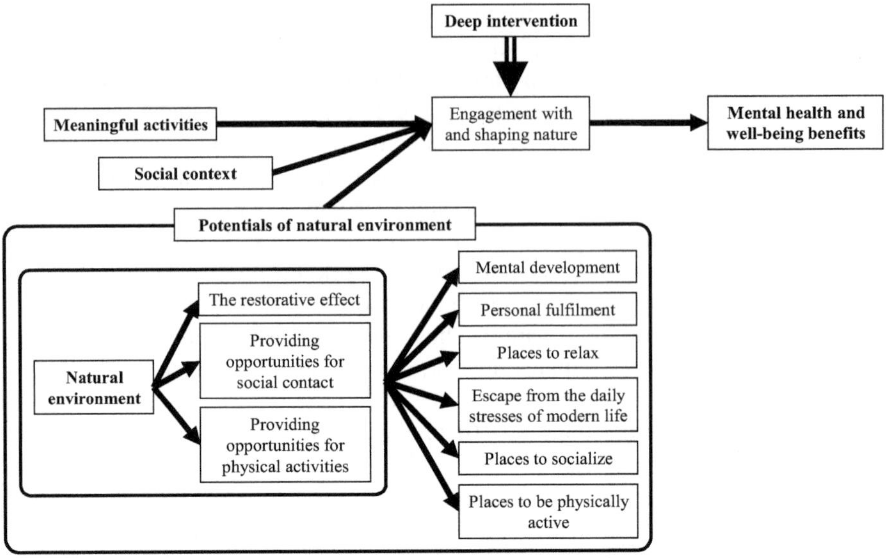

Fig. 6.2 Logic model that could be developed by and included in Bragg and Atkins (2016) regarding the manner mental health and well-being benefits derive from care/social farming

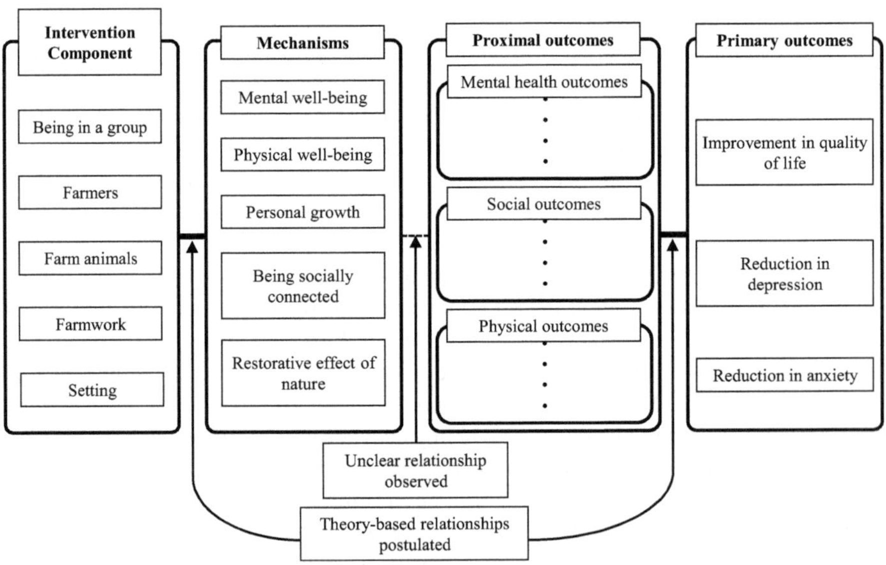

Fig. 6.3 General framework of logic models developed by Murray et al. (2019a) describing how care/social farming works for different population groups

discussed concept of the "restorative effects of nature," although "mental well-being" was frequently referenced. The authors interpret this as an indication that the mechanisms interact in an unknown way, a common feature of logic models. Nevertheless, the article emphasizes that nature remains the essential intervention component that activates other, more observable mechanisms.

Once again, I identify a "black box" obscuring the links between nature and mental health, mirroring the issue I discussed in Chap. 2 regarding the relationship between physical strength and the sense of liberation that arises from witnessing the "lawful" nature of non-human things. I intentionally distinguish the orderly aspect of nature from the general concept of nature found in rural environments, attributing the therapeutic properties of farming to the former. Gorman (2020) notes that rural landscapes can both foster and hinder therapeutic possibilities; for example, the sight of deceased farm animals may disrupt therapeutic processes. Conversely, I view even the observation of death as a therapeutic natural phenomenon. The laws of nature, while sometimes harsh, can be emancipatory, offering therapeutic value by exposing individuals to the realities of the natural world.

Murray et al. (2019b) aimed to explore whether and how care farming could support the rehabilitation of probation service users under community orders. Despite highlighting the significant role of farmers in facilitating interactions with animals and farmwork, a newer logic model was found to share the same structure as earlier ones. Payne et al. (2023) revisited the literature on the rehabilitative potential of farm animals in probation services, concluding that it remains under-researched. In the context of nature-based interventions for vulnerable youth, Overbey et al. (2023) revealed that animal-assisted therapy has yielded more robust findings, largely due to its narrower definition compared to other types of nature-related interventions, such as wilderness therapy, horticultural therapy, and care farming. The authors noted that care farming has contributed most significantly to mental health, social relationships, and personal development. Care farm activities involve a wider range of tasks, though farm animals tend to attract more attention from researchers and practitioners than crops.

Thus far, the logic models for care/social farming mentioned above have not been updated by subsequent studies. Consequently, I regard them as the most recent attempt to explain the mechanisms of well-being enhancement. Unlike the theory of autonomous horticulture proposed in the previous chapter, the logic models are based on empirical research, including interviews, without a theoretical focus on specific components of care farming settings. These models encompass various elements that are not exclusive to farming. For example, there is debate about whether non-farm workplaces can offer benefits comparable to those derived from interaction with farmers and participation in farmwork. Factory supervisors, like care farmers, can be sympathetic toward clients, and factory work can provide as much physical exercise as farm work. Furthermore, the liberation mechanism identified in Chap. 2 could also operate in non-farm workplaces, as the law of nature governs both living and inanimate matters, limiting the arbitrariness of superiors at work. The key difference lies in the individuality and uncertainty inherent in life. The logic models appear designed to capture all potential advantages of care farming over conventional care provision, rather than focusing exclusively on its benefits compared to other therapeutic occupations. This explains why discussions often drift away from the unique aspects of farming.

Regarding the "black box" (cf. Fig. 6.1), the logic models suggest that different population groups may benefit from the same multifaceted activity in diverse ways.

While it is natural for people to derive different benefits from the same activity, the critical question is whether these varied benefits stem from a single aspect of farming with interconnected cause-and-effect relationships, or from several aspects independently producing distinct outcomes. The black box complicates this issue. If the former is true, the multifaceted nature may be superficial; if the latter is true, service providers can disassemble a care farm to select the most useful components for each client group. In either case, research could have specified the multiple benefits of care/social farming through a priori speculation, rather than relying entirely on empirical data intended to identify every aspect responsible for meeting specific needs. This alternative research strategy aligns with a social work practice approach from a "strength-based perspective" (Berg-Weger 2019). Correspondingly, Fieldhouse (2003) emphasized that services should focus on clients' skills and aspirations, rather than their symptoms and deficits, to integrate clients into broader communities rather than segregating them. This presents a contrast between evidence-seeking fragmented studies and earlier, more holistic approaches.

6.3.3 Why Gardeners Garden?

Norwood (2022) raises a broader question about why people garden. The author suggests that, while the evolutionary explanation of gardening, linked to the restorative properties of nature, may be technically accurate, it does not capture the deeper human significance of gardens. Thus, a philosophical approach is adopted, asserting that gardening, particularly through cultivating domesticated plant species, serves as an intellectual quest that ultimately teaches us how to live. Gardening is presented as an activity that engages the mind, body, and spirit. As Norwood (2022) writes: "Mother Nature never ceases presenting obstacles; one cannot be a successful gardener without continually learning new things, and that too is part of [a] good life: always asking questions, finding answers, developing new questions, and never achieving perfection, but pursuing it anyhow." This challenge, when overcome, leads gardeners to a state of elegance where they can "acquire not only basic physiological needs such as food and exercise but also aesthetic experiences." This aligns with the difficulty-conscious approach in the current chapter and resonates with the words of a former professional farmer, cited in Chap. 4, Case B1.

Similar to the earlier reference to Waliczek et al. (2005), Norwood (2022) also supports the notion that "the garden is one of the few things able to help humans obtain every level of Maslow's hierarchy of needs simultaneously." The benefits of gardening for gardeners extend across multiple aspects, affecting various levels of Maslow's hierarchy. Even when gardeners grow crops solely for the end products, they inevitably expand their knowledge to ensure success throughout the current cropping season. This process brings by-products such as improved preparedness for future contingencies, enhanced flexibility, opportunities to exchange cultivation techniques with colleagues, and a greater appreciation for the value of their knowledge. These outcomes fulfill needs beyond the physiological, addressing higher-level needs in Maslow's hierarchy. As Norwood (2022) illustrates, successful knowledge acquisition in gardening contributes to the fulfillment of "belonging," "esteem," "cognitive," "aesthetic," and "self-actualization" needs. This establishes a causal sequence across different hierarchical levels, each related to various aspects of gardening (Fig. 6.4). The author provides a more comprehensive explanation of meaningfulness than the logic models used in care/social farming. Therefore, growing crops can account for a wide range of outcomes, which justifies strength-based approaches for communities with diverse needs.

This discussion draws from the thoughts of several historical thinkers, particularly David Cooper, who examined how gardening addresses what Norwood refers to as "transcendental needs," which occupy the highest levels of the hierarchy. Cooper (2006) distinguishes between "garden-practice" and "gardening," as the latter can simply involve the contemplation of plants. Cooper's primary argument is that garden-practice induces virtues such as care and respect, self-mastery and self-discipline, and humility and hope—all of which contribute to "unselfing." Gardeners must approach plants with care and respect, submitting to natural structures and patterns as a means of achieving self-discipline. Their belief that farm produce results not just from labor but from grace reflects their humility, which is rooted in hope and trust in the cooperation of the world. Additionally, those who understand co-dependence recognize that human creativity depends on non-human beings and is essential for creating a meaningful world. Finally, gardening cultivates virtues of serenity, peace of mind, and satisfaction derived from living in truth.

In this context, "truth" refers to the "directions" gardeners receive from their crops, and many of the aforementioned virtues correspond to the outcomes of the I-did-it experience. Cooper concludes that there is happiness in caring for living things in response to their needs. However, he qualifies this statement, arguing that garden-practice is concerned only with truth, not with happiness, if happiness is

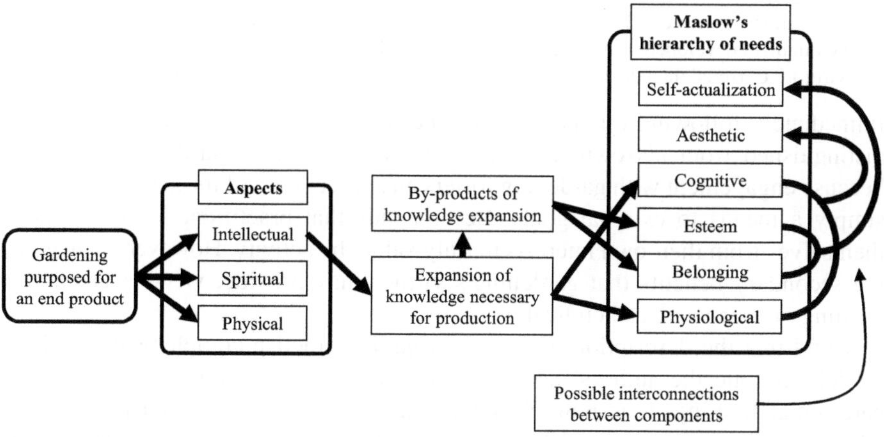

Fig. 6.4 Relationships between the intellectual aspect of gardening and every level of need in Maslow's hierarchy, an illustration based on Norwood (2022)

understood in the modern sense—primarily as pleasure or contentment (Cooper 2006, Location 1858/1996). This claim critiques modern notions of happiness, which often reflect a desire for greater autonomy from nature's laws. Gardening, in contrast, invites reflection on whether contemporary definitions of happiness are an appropriate standard for a fulfilling life. Despite critiques of happiness, the concept has been rigorously examined within the context of care/social farming (cf. de Moor and Hermsen 2018).

However, a key issue remains: both Cooper and Norwood focus on autonomous gardeners, seemingly distancing themselves from the use of horticultural activities for external purposes. Their arguments implicitly reference care/social farming or other nature-based interventions but do so in a way that limits their application to autonomous gardening practices. Consider the following quotations:

> There is indeed something about gardening that appeals to human biological functions, suggesting the desire is innate—so much so that it is a successful treatment for mental health issues. However, it is unclear exactly which aspects of gardens improve biological functioning: the exercise, interaction with plants, fresh air, or all of the above? (Norwood 2022: 305)

The uncertainty expressed here mirrors the earlier analogy of the logic models as a black box. Originating from empirical studies, these models cannot account for all the aspects present in the data, even those unrelated to care farming. As a result, the models may appear more heterogeneous than those constructed using a "philosophical approach." One reason for this discrepancy is that the inclusion of elements unrelated to gardening obscures the causal relationships among various gardening aspects. However, excluding these elements may lead to the omission of important factors. Therefore, the present volume integrates an interview-based empirical approach with a deductive framework to link the I-did-it experience to autonomous gardening. Another quotation highlights a similar issue:

> … here are three views found in this tradition that should be distinguished from my claim. First, there's the idea that simply being in gardens, surrounded by plants, enhances people's feeling of well-being, especially those who are ill or depressed, and thereby enables them to live more effectively. Then, … gardening stiffens resistance to temptation. … Finally, … the qualities developed through gardening … would help to produce more responsible and 'virtuous citizens' than those brought up in the cities … (Cooper 2006: Location 1085/1996)

Immediately following this passage, Cooper explains why these views should be distinguished from his own. Certain garden-practices necessarily induce virtues because engagement with gardening fosters an appreciation of the activity itself, not simply a means to external goals. He contends that practitioners must distance themselves from their own interests to truly value the activity. However, he neglects the secondary benefits that gardening's virtues may provide, viewing care/social farming or green care as heretical.

I find that the description of virtues in garden-practice complements the logic models, despite the authors' avoidance of explicitly associating their work with care/social farming. It is now clear that the "philosophical approach" offers an advantage over empirical models: it can reveal the logical connections among components in the "black box." This ability will help further clarify the present attainments of research on care/social farming.

6.3.4 Synthesis of the Present Research on Care/Social Farming

I will attempt to synthesize the current research on care/social farming here. First, care/social farming practitioners are the figures that use the virtues of gardening to benefit their clients who participate in as non-autonomous gardeners. Regarding the role of farmers in care/social farms, it is logical to consider that the "interaction" between care farmers and participants mirrors the "instruction giver-receiver relationship." As described in Chap. 3, during farmwork, these parties communicate "directions" derived from crops, which are in harmony with the laws of nature. In doing so, care farmers share their experiences with clients. If Cooper's (2006) description applies to care farmers, this interaction embodies the virtues fostered by gardening practices. In this sense, care farmers "unself" themselves to embody nature before their clients.

Recent models of care/social farming prominently feature "interactions with farm animals," as several studies emphasize the distinct role animals play for clients (Hassink et al. 2017; Sudmann and Børsheim 2017; Gorman and Cacciatore 2022; Thieleman et al. 2023). Gorman (2019) contends that such therapeutic practices can be unjustly anthropocentric and utilitarian. However, as Leck et al. (2014) report from the UK, care farmers recognize that many clients prefer engaging with livestock over crops. Nonetheless, the primary motivation for care farmers in the UK remains their ability to make a tangible difference in vulnerable people's lives. A variety of animal species are often present in care farms, distinguishing them from industrial farms focused on profitability. Care/social farming is thus presented as an antithesis to the neoliberal agrarian regime and utilitarian productivism. Some practitioners even reject the term "care farming," which they view as stigmatizing and patronizing in the context of mainstream care practices, proposing "connective agriculture" instead. Care farms enable both staff and clients to live according to their true selves (Hemingway et al. 2016). In relation to the logic models, the focus has shifted from "interactions with farm animals" to "interactions with farmers." The question is not whether animals or plants contribute more significantly to the therapeutic process but what care farmers' criticism targets.

Hassink et al. (2014a, b) note that the pioneers of care farming in the Netherlands were critical of both mainstream agriculture and care practices. They sought alternatives to agricultural intensification while addressing environmental concerns, advocating for the normalization and socialization of clients. These farmers focused on participants' individual potential rather than their limitations. As de Moor and Hermsen (2018) observe, participants often experience a "farmer and co-worker" relationship in addition to, or instead of, the "caregiver and client" dependency, evaluating care farmers' personal approach, attention, and involvement with the participants' well-being. Without an overemphasis on disabilities or psychiatric histories, these farmers seem to have already gained the strength-based perspective. Similarly, Swedish care farmers critique both production-oriented farming and diagnosis-based care systems, shifting their focus to cultivating connections, not animals and plants anymore. Practitioners even stressed that participants should be on the farms to care for others, rather than being cared for (Pettersson and Tillmar 2022).

The critiques of mainstream agriculture offered by these authors center on the dualism between production and care, though I interpret this differently. I propose the concept of "true productivism," which holds that sustainable production of both living and non-living materials must align with the principles of care. Production and care, in their truest sense, are not opposites but are inherently compatible. The misconception of care as counterproductive arises from the shortsightedness of capitalist projects. It is unsurprising that the labor market has deviated from the natural laws in response to the demands of capitalism. Care farmers, with their deep understanding of various modes of agriculture, are better suited as instructors in fostering harmony between production and care.

Hudcová (2022) acknowledges the advantages of farming activities, which include farmers' respectful, person-centered approach, the inclusive nature of the farms, and the rural environment itself. However, she questions whether these advantages translate into "empowerment," which she defines as the opposite of powerlessness, submission, and oppression. Empowerment-based social work practices, she argues, should involve communal actions and the redistribution of power to challenge oppressive systems that marginalize vulnerable groups. Current literature on care/social farming, however, primarily focuses on promoting personal competence in clients with mental disabilities, rather than addressing power dynamics within a broader societal context. The connection between farming and social reform lies in addressing the problems created by mainstream society. The author acknowledges that research on this topic is still emerging, yet there is a logical gap that needs to be bridged with insights from social work, the largest profession within the social welfare system (Abramovitz 2014).

6.3.5 Empowerment and Social Work

Payne (2021) highlights that the social construction approach, when applied to social work, enables practitioners to influence both social institutions and individuals. This process is gradual, with changes becoming apparent only after they have accumulated significantly. Social construction occurs through social interactions, wherein individuals use language to create shared perceptions of reality. In social worker-client relationships, this process can deconstruct previously accepted truths and generate alternative ways of thinking. This aligns with the postmodernist argument that "we should not take the present social order for granted." By recognizing these new perspectives, clients can understand that they may have unfairly blamed themselves for their circumstances. Social workers, in turn, can empower clients, helping them regain a sense of control over their lives and the events that affect them. This process of empowerment is central to progressive social work practices, though it does not necessarily aim for broad social change. Such ambivalence also aligns with traditional social work approaches, which tend to reduce complex social problems to individual psychological issues, reinforcing the idea that individuals should adapt to the current social order rather than question or resist it.

The critical perspective on traditional social work intersects with the views of care farmers, who seek alternatives to mainstream care practices, as discussed in the previous subsection. Payne (2021) also raises concerns about the professionalization of social work, which prioritizes technical skills over moral and political aspects of helping. This tendency would serve to distinguish social work from other professions, promoting professional hierarchies. Instead, social workers could form alliances with non-professional groups, such as working-class and community organizations, rather than exclusively with professional entities. Several studies on care/social farming indicate that farmers, despite not being health or caregiving professionals, have successfully challenged this professionalization and achieved positive intervention outcomes, earning appreciation from both professionals and participants (e.g., Hassink et al. 2014a, b; Schreuder et al. 2014; Lund et al. 2015). If political awareness and personal activism are considered more critical than social work techniques, as Lund et al. (2015) suggest, amateurism or anti-professionalism, fueled by enthusiasm, will be responsible to counterbalance traditional healthcare models. Meanwhile, professionalism remains embedded within the cultural and social systems of countries or localities, allowing social work practices to change through their impact on global discourses. This underscores the ongoing importance of professionalization in social work (Payne 2006: 190). Critical social work could foster new relationships with non-professional carers, which would open up additional avenues for intervention.

One theory suggests that, while addressing the personal consequences of oppression for individuals is essential, practitioners must also focus on oppressors—those benefiting from current oppressive systems. These individuals need assistance in understanding and accepting personal responsibility for their actions, even if such behavior is often unintentional, stemming from structural factors (Payne 2021). Further scholarship addresses how social workers should respond to privilege—the counterpart of oppression—by working both within and outside the system, with a focus on political action. Social workers must confront issues of oppression and privilege, which are inseparable and thrive on silence. Such actions would empower marginalized communities by addressing discrimination and oppression.

Theories of social work that incorporate garden activities are still limited in this regard, though some scholars suggest using farmwork or community gardening as tools for intervention (cf. Dominelli 2012; Mama 2018; Payne 2021). Dominelli (2012: Location 1731/5758) provides an example of how gardening can empower a disinvested community that is "alienated from society and isolated from its wider networks. A community worker, raised in a more affluent area where his mother cultivated a garden, sought to share this experience with residents. Over time, he inspired them to create "a space for enacting solidarity, and caring for, and being cared by, others" (Dominelli 2012: Location 1782/5758). He acted as a catalyst for empowering individuals and helping them tap into their resilience, preparing them to confront structural inequalities. This example nearly achieves empowerment, as the worker shared his personal background and his gardening experience with an awareness of his privilege. However, existing literature does not yet address how collective gardening activities might challenge the dominance of oppressive societal groups. A comprehensive conclusion will address this issue in the remainder of the chapter.

6.4 Conclusion

The final task is to compare this volume with existing literature. The concluding section should identify an oppressive force that justifies the need for the agriculture-welfare partnership or care/social farming, while also proposing an alternative. To illustrate this, we revisit the concept of "meaningfulness." The term is often implicitly associated with spirituality or connectedness to creation (cf. Payne 2021), as seen in some of the studies referenced earlier. Meanwhile, the meaningfulness of work or life has been discussed within business administration and philosophy (e.g., Rosso et al. 2010; Thomas 2019). However, the precise meaning of "meaningfulness" has not been sufficiently examined. A specific explanation emerged only with the development of the logic models, which states:

> Service users perceive tasks as meaningful because they are judged to be useful to others and are needed to conduct day-to-day activities at the farm. Service users also see their role as personally meaningful, contributing to society giving them a sense of purpose, happiness and fulfilment (Murray et al. 2019a: 36).

It is not surprising that the concept of meaningfulness has evolved beyond what a single model can capture, even though it is classified as an explanatory factor of "personal growth" among "mechanisms" in the models under consideration (Fig. 6.3). Furthermore, beyond farmwork or care farms, there are other occupations or sheltered workshops where service users may experience similar positive emotions. The explanation above remains a general claim that care farming is meaningful or merely a positive assessment of it. Meaningfulness becomes apparent only when contrasted with the meaninglessness of something else. A pertinent example is the concept of "bullshit jobs," which has been described as "spiritual violence" (Graber 2018); I would argue that it reflects underlying power relations in workplaces. Additionally, as Murray et al. (2019b) note, farmwork is sometimes viewed as punishment rather than rehabilitation within the probation context. These authors suggest a conflict between the "philosophies of care farming" and probation service principles, which may undermine the true value of care farming for offenders.

This distortion is not unique to the present context; it corresponds to the ascetic attitude toward labor discussed in earlier chapters. I have argued that as long as laborers deny themselves the pleasure of growing crops, they can never enjoy the I-did-it experience, let alone improve their skills in observation of crop growth. This connects with the vilification of laborers, as previously mentioned. For those excluded from the mainstream labor market, one way to avoid such vilification is to actively endure punishments in anticipation of satisfying mainstream society's expectations. Thus, farm tasks may need to appear punitive to those potential critics lurking in the public's view. Both laborers in despair and their critics share the notion that work is essentially a form of punishment, which, according to this belief, disciplines workers. Likewise, job seekers and welfare recipients may be seen as needing discipline, provided by a substitute for traditional employment. Graber (2018) also suggests that this perspective is widespread, contributing to the proliferation of "bullshit jobs."

The previous discussion offers a counterargument: it is not punishment but pleasure that enables one to become a skilled grower or carer of life. By replacing the prevailing notion, this thesis aims to neutralize the slanders and, in turn, motivate not only those at risk of self-punishment but also the critics who cling to the mainstream economy by overfitting themselves to maltreatment in their workplaces. Trainees on farms will then realize that their joy is worthwhile, and those "privileged" workers will finally recognize that they lack such, if not delight, meaning in their jobs. The difficulty-conscious approach maintained in this volume differs from the lighthearted self-punishment attitude displayed by laborers who risk slander, and so does the concept of "self-discipline" noted in Cooper (2006). It should be remembered that domesticated life forms have sustained their lives by providing aesthetic experiences to growers, although this evolutionary strategy has largely lost its relevance in today's unprecedentedly urbanized society. Urbanization and the condemnation of work-related pleasure follow similar trajectories. In other words, when aesthetics is integrated into wage labor, the norms we have come to accept in the existing labor market cannot remain unchanged.

Therefore, horticulture, if properly conveyed, is not only therapeutic for welfare beneficiaries but for society as a whole. This therapeutic aspect was likely evident in early studies, which referred to "connectedness with nature" (cf. Sempik et al. 2005). However, research could have more explicitly described this as "emancipation from the arbitrariness of human superiors," with the understanding that people are ultimately equal before the laws of nature. In addition, if research had dealt challenges faced in horticultural activities, some equivalent of the I-did-it experience could have been provided with a more nuanced description.

The preceding discussion demonstrates how social discourses shape the success or failure of novice horticulturists in acquiring autonomy through the instructor-participant relationship. In this context, instructors and participants may need to collaborate with social workers to present a counterargument. Progressive social work practices, which have contributed to anti-discrimination movements, could be instrumental in communicating the intrinsic value of horticulture. By doing so, practitioners can even appeal to the "privileged" class, who conventional social work would not approach. Empowerment will thus occur in the context of care/social farming or the agriculture-welfare partnership. It may seem that I am excessively concerned about the re-exclusion of farm participants from citizenship, but this preventive countermeasure ultimately proves constructive, as it integrates critical theories of capitalism and social work to introduce the new concept of "employment refusal support" as an alternative to traditional employment support. Why agriculture for social welfare? Because it reminds us that work done for the care of life, even if performed for wages, must be delightful.

Throughout this chapter, I have defended the difficulty-conscious approach. Now, it is appropriate to elaborate on why experiencing the joy of horticulture is difficult for some individuals. A theory of autism may furnish insight. Murakami (2008) argues that children with autism exist in a sphere where no living thing attempts to associate with them, meaning they do not need to distinguish between living and non-living things. However, once they make eye contact with someone, they unexpectedly perceive themselves as entering a new dimension. This shifts their perspective on living things. Murakami applies this phenomenon to the way children with autism develop interpersonal relationships. Similarly, many adults "without autism" face challenges in forming relationships with plants. Autism toward humans does not equate to autism toward plants, and vice versa. Establishing relationships with plants is as difficult for non-autistic adults as developing interpersonal relationships is for autistic children. If you appreciate the growth of a plant, you recognize that the plant is an irreplaceable living being—proof that you have overcome your "autism" toward plants. What a meaningful miracle!

References[3]

Abramovitz M (2014) Definition and functions of social welfare policy: setting the stage for social change. In: Blau J (ed) The dynamics of social welfare policy, 4th edn. Oxford University Press, New York, pp 19–63

Berg-Weger M (2019) Social work and social welfare: an invitation, 5th edn. Routledge, New York

Bragg R, Atkins G (2016) A review of nature-based interventions for mental health care. Natural England Commissioned Reports, No. 204

Chakrabarti, Anjan, Stephen Cullenberg and Anup Dhar (2017). Primitive accumulation and historical inevitability: a postcolonial critique. In: Theodore Burczak, Robert Garnett Jr and Richard McIntyre (eds.) Knowledge, class, and economics: marxism without guarantees, Routledge: London, pp. 288–306

Cooper DE (2006) A philosophy of gardens. Oxford University Press Inc, New York

Crinson I, Yuill C (2008) What can alienation theory contribute to an understanding of social inequalities in health? Int J Health Serv 38(3):455–470

Csikszentmihalyi M (1992) Flow: the psychology of happiness. Random House, London

de Moor D, Hermsen M (2018) Achieving happiness at care farms in the Netherlands. J Social Intervent Theory Pract 27(6):4–23

Dominelli L (2012) Green social work: from environmental crises to environmental justice. Polity Press, Cambridge

Elings M (2006) People-plant interaction. In: Hassink J, van Dijk M (eds) Farming for health: green-care farming across Europe and the United States of America. Springer, pp 34–55

Fieldhouse J (2003) The impact of an allotment group on mental health: clients' health, wellbeing and social networking. Br J Occup Ther 66(7):286–296

[3] ***: in Japanese, **: in Japanese with an English title, *: in Japanese with an English title and summary.

Fraser N (2022) Cannibal capitalism: how our system is devouring democracy, care, and the planet—and what we can do about it. Verso, London

Glassman J (2006) Primitive accumulation, accumulation by dispossession, accumulation by 'extra-economic' means. Prog Hum Geogr 30(5):608–625

Gonzalez MT, Hartig T, Patil G, Martinsen EW (2011) A prospective study of existential issues in therapeutic horticulture for clinical depression. Ment Health Nurs 32:73–81

Gorman R (2019) What's in it for the animals? Symbiotically considering 'therapeutic' human-animal relations within spaces and practices of care farming. Med Humanit 45:313–325

Gorman R (2020) Care farming, therapeutic landscapes, and rurality in the UK. In: Munoz S, Bain S (eds) Mental health and wellbeing in rural regions: international perspectives. Routledge, London, pp 133–150

Gorman R, Cacciatore J (2022) Exploring the effects of the human-animal relationship on care-farms in the context of trauma histories. Anthrozoös 36(2):163–177

Graber D (2018) Bullshit jobs. Simon and Schuster, Inc., New York

***Hamada K (2018) The present welfare-oriented agriculture, the problems confronting it and future prospects [Fukushinōgyō no genjō, kadai, tenbō]. Agric Hortic 93(9):793–798

Hardman M, Larkham PJ (2014) Informal urban agriculture: the secret lives of guerilla gardeners. Springer

Hassink J, van Dijk M (2006) Farming for health across Europe. In: Hassink J, van Dijk M (eds) Farming for health: green-care farming across Europe and the United States of America. Springer, pp 347–357

Hassink J, Grin J, Hulsink W (2014a) New practice of farm-based community-oriented social care services in the Netherlands. J Soc Serv Res 0:1–15

Hassink J, Hulsink W, Grin J (2014b) Farming with care: the evolution of care farming in the Netherlands. Wageningen J Life Sci 68:1–11

Hassink J, De Bruin SR, Berget B, Elings M (2017) Exploring the role of farm animals in providing care at care farms. Animals 7(45). https://doi.org/10.3390/ani7060045

Hayashi R, Okuhira M (2001) The Disability Rights Movement in Japan: past, present and future. Disabil Soc 16(6):855–869. https://doi.org/10.1080/09687590120083994

He S, Li M, Zheng Y, Gao X, Wang R (2024) Reflections on the production of space and justice: a study of everyday life of the SuoJincun's guerrilla gardeners. Cities 154:15345

Hemingway A, Ellis-Hill C, Norton E (2016) What does care farming provide for clients? The views of care farm staff. NJAS Wageningen J Life Sci 79:23–29

Hirono S (2014) The backgrounds of conflicting discourses and activism concerning Fuchu Center for Therapy and Education. Joint J Natl Univ Kyushu Educ Humanit 2(1):10

Hudcová E (2022) Social work in social farming in the concept of empowerment. Eur Countryside 14(3):497–510

Hung H (2017) Formation of new property rights on government land through informal co-management: case studies on countryside guerilla gardening. Land Use Policy 63:381–393

Iancu SC, Hoogendoorn AW, Zweekhorst MBM, Dick JV, Bunders JFG, van Balkom AJLM (2013) Farm-based interventions for people with mental disorders: a systematic review of literature. Disabil Rehabil 37(5):379–388

***Ikegami K (2013) The potential of agriculture for welfare: an attempt at agromedicopolis [Nō no Fukushiryoku: Agro-mediko-porisu no chōsen]. Rural Culture Association Japan, Tokyo

Irvine KN, Warber SL (2002) Greening healthcare: practicing as if the natural environment really mattered. Altern Ther Health Med 8(5):76–83

Jaffe JA, Gertler M (2006) Victual vicissitudes: consumer deskilling and the (gendered) transformation of food systems. Agric Hum Values 23:143–162

***Kanda H, Maiko N, Naoko H, Harumi T, Yasunori Y, Katsunori K (2001) A report on opinion polls among welfare and medical institutions in Akita Prefecture regarding horticultural therapy [Akitakennai no fukushishisetsutō ni okeru engēryōhō ni kansuru ishikichōsahōkokusho]. J Akita Prefectural Junior Coll 2:13–21

Kaplan S (1995) The restorative benefits of nature: toward an integrative framework. J Environ Psychol 15:169–182

Kaplan R, Kaplan S (2011) Well-being, reasonableness, and the natural environment. Appl Psychol Health Well Being 3(3):304–321

Leck C, Evans N, Upton D (2014) Agriculture—who cares? An investigation of 'care farming' in the UK. J Rural Stud 34:313–325

Lund IE, Granerud A, Eriksson BG (2015) Green care from the provider's perspective: an insecure position facing different social worlds. SAGE Open 5. https://doi.org/10.1177/2158244014568422

Mama RS (2018) Community gardening: the nexus for community, social work and university collaboration. In: Dominelli L (ed) The Routledge handbook of green social work. Routledge, New York, pp 182–191

***Matsumoto H (2014) Voluntariness of social welfare: a Commentary for Shigeo Okamura's "The principles of social welfare" [Shutaisē no Shakaifukushi: Okamura Shigeo cho "Shakaifukushigenron" o toku]. Sangaku Publishing, Otsu, Japan

Matsuo E (1998) What is "horticultural wellbeing" in relation to "horticultural therapy"? In: Burchett MD, Tarran J, Wood R (eds) Towards a new millennium in people-plant relationships. University of Technology, Sydney Printing Services, Sydney, pp 174–180

***Matsuo Ē (2005) An exhortation to horticulture: environment, education, welfare and community development. Rural Culture Association Japan, Tokyo

***Murakami Y (2008) The phenomenology of autism [Jihēshō no Genshōgaku]. Keiso Shob, Tokyo

Murray J, Wickramasekera N, Elings M, Bragg R, Brennan C, Richardson Z, Wright J, Llorente MG, Cade J, Shickle D, Tubeuf S, Elsey H (2019a) The impact of care farms on quality of life, depression and anxiety among different population groups: a systematic review. Campbell Systemat Rev 15(4):e1061

Murray J, Cokerb JF, Elseya H (2019b) Care farming: rehabilitation or punishment? A qualitative exploration of the use of care farming within community orders. Health Place 58:102156

***Nakamura H (2002) The pursuit of concernedness and participatory development: social actions of universities in Sri Lanka [Tōjishasē no tankyū to sankagatakaihatsu: Suriranka ni miru daigaku no shakaikōkenkatsudō]. In: Saito F (ed) Participatory development: to empower disadvantaged people to take initiatives in developmental projects [Sankagatakaihatsu: Mazushī Hitobito ga Shuyaku to naru Kaihatsu he mukete]. Nippon Hyoron sha, Tokyo, Japan

***National Institute for Rural Engineering (2008) A guide to employment of people with disabilities in the agricultural sector: case examples [Nōgyōbun-ya ni okeru Shōgaishashūrō no Tebiki: Sagyōjirēhen]. National Institute for Rural Engineering

Nichols R (2015) Disaggregating primitive accumulation. Radic Philos 194:18–28

Norwood FB (2022) The garden as art, hobby, and the good life. Horttechnology 32(3):304–312

Ochiai C (2014) Traditional community-based disaster management in World Heritage Site of Shirakawa Village. In: Shaw R (ed) Community practices for disaster risk reduction in Japan. Springer Japan, pp 155–173

**Okada T (2022) Conflict over mutual support in the "Okamura Theory": focusing on the concept of "Welfare as a Folklore". J Tokyo Online Univ 4:47–62

*Ootou F (2014) Positioning of the community in the theory of Shigeo Okamura. Social Inf Sci 19:1–12

Overbey AT, Diekmann F, Lekies KS (2023) Nature-based interventions for vulnerable youth: a scoping review. Int J Environ Health Res 33(1):15–53

Oversveen E (2022) Capitalism and alienation: towards a Marxist theory of alienation for the 21st century. Eur J Soc Theory 25(3):440–457

Payne M (2006) What is professional social work? Revised, 2nd edn. Policy Press, Bristol

Payne M (2021) Modern social work theory. Bloomsbury Publishing, London

Payne L, McMurran M, Glennan C, Mercer J (2023) The impact of working with farm animals on people with offending histories: a scoping review. Int J Offender Ther Comp Criminol 67(12):1282–1302
Perelman M (2000) The invention of capitalism: classical political economy and the secret history of primitive accumulation. Duku University Press, Durham
Perelman M (2007) Primitive accumulation from feudalism to neoliberalism. Capital Nat Social 18(2):44–61. https://doi.org/10.1080/10455750701366410
Pettersson K, Tillmar M (2022) Working from the heart: cultivating feminist care ethics through care farming in Sweden. Gend Place Cult 29(10):1446–1466
Rionx S, LeBaron G, Verovsek PJ (2019) Capitalism and unfree labour: a revire of Marxist perspectives on modern slavery. Rev Int Polit Econ 27(3):709–731
Rosner AL (2012) Evidence-based medicine: revisiting the pyramid of priorities. J Bodyw Mov Ther 16(1):42–49
Rosso BD, Dekas KH, Wrzesniewski A (2010) On the meaning of work: a theoretical integration and review. Res Organized Behav 30:91–127
Schreuder E, Rijnders M, Vaandrager L, Hassink J, Enders-Slegers M-J, Kennedy L (2014) Exploring salutogenic mechanisms of an outdoor experiential learning programme on youth care farms in the Nertherlands: untapped potential? Int J Adolesc Youth 19(2):139–152
Sempik J, Aldridge J, Becker S (2005) Health, well-being and social inclusion: therapeutic horticulture in the UK. Policy Press, Bristol
Sudmann TT, Børsheim IT (2017) 'It's good to be useful': activity provision on green care farms in Norway for people living with dementia. Int Pract Dev J 7(Special Issue on Enhancing wellbeing: practice and politics):1–14
Suehara T (2006) Labor exchange systems in Japan and DR Congo: similarities and difference. Afr Stud Q 9(1&2):55–65
Sugimoto Y (2015) An introduction to Japanese Society, 4th edn. Cambridge University Press
**Takeyama U (2019) From the horticultural well-being to the development of social resources: case of Yamashina Ward, Kyoto City. Komazawa J Sociol 52:23–54
Thieleman K, Cacciatore J, Gorman R (2023) "Perhaps something of beauty can grow:" experiences of care farming for grief. Death Stud 46(10):2435–2444
Thomas JL (2019) Meaningfulness as sensefulness. Philosophia 47:1555–1577
Thompson CW, Roe J, Aspinall P, Mitchell R, Clow A, Miller D (2012) More green space is linked to less stress in deprived communities: evidence from salivary cortisol patterns. Landsc Urban Plan 105(3):221–229
Trombly CA (1995) Occupation: purposefulness and meaningfulness as therapeutic mechanisms. Am J Occup Ther 49(10):960–972
Ulrich RS (1983) Aesthetic and affective response to natural environment. In: Altman I, Wohlwill J (eds) Human behavior and environment, Behavior and natural envirnment, vol 6. Plenum, New York, pp 85–125
Ulrich RS, Parsons R (1992) Influences of passive experiences with plants on individual well-being and health. In: Relf D (ed) The role of horticulture in human well-being and social development: a National Symposium, April 19–21, 1990, Arlington, Virginia. Timber Press, Portland, pp 93–105
Unruh AM (2004) The meaning of gardens and gardening in daily life: a comparison between gardeners with serious health problems and healthy participants. Acta Hortic 639:67–73
Waliczek TM, Zajicek JM, Lineberger RD (2005) The influence of gardening activities on consumer perceptions of life satisfaction. HortScience 40(5):1360–1365

Chapter 7
Conclusion

Abstract This chapter begins by summarizing key points from earlier discussions, including a feedback comment received from an attendee at a university lecture. The comment relates to the "difficulty-conscious approach" introduced in earlier chapters and includes a few offensive remarks about homeless people. In my lecture, I aimed to highlight both the challenges inherent in farming and the significant potential of farm participants, acknowledging a trade-off between the two. My ongoing efforts to resolve the self-contradiction revisited central themes throughout the book. After briefly addressing the question posed in the first chapter, the latter half explores additional discussions that could emerge. Research on the agriculture-welfare partnership and care/social farming exemplifies the multifunctionality of agriculture, showing how its externalities can lead to substantial changes in social work practices. The movement toward integrating agriculture and social welfare seeks to reinternalize externalities previously excluded from research by Japanese agronomists for specific reasons. The chapter concludes by questioning the attitudes of actors in other industries and revisiting the concept of "social inclusion," which assumes a new significance.

Keywords Agriculture-first principle · Discriminatory remark · External economy · Pedagogy · Social work

7.1 A Pedagogic Issue Coming Up in Academia

This volume initially emerged from my simple aspiration to create alternative employment opportunities for homeless individuals. However, I soon discovered that realizing this goal was not straightforward, prompting me to theorize the challenges involved. Although the original objective remains unmet, the action research project has provided valuable insights into why horticultural activities remain beneficial from the perspective of social welfare and reform, as demonstrated in previous chapters. Additionally, the preceding section emphasized that engaging mainstream citizens could significantly influence the future of such efforts.

H. Tsunashima, *Farming and Social Inclusion in Japan*, Urban Agriculture,
https://doi.org/10.1007/978-3-032-01598-3_7

This chapter concludes with an episode from my experience as a university lecturer. As I progressed with the action research project, I was invited to give guest lectures in a university course on social activities. Given the broad scope of the course, the organizers expected me to present my project as a case of agriculture-welfare collaboration. After each lecture, I welcomed critical feedback from attendees, typically in written form. However, because the course covered diverse topics, not all students were interested in or prepared to engage with issues of social exclusion. On one occasion, I encountered a particularly discriminatory comment:

> [...] Those farmwork opportunities may indeed provide homeless laborers with experiences in *real society*, but I do not want them to purpose that. Not a few people would have a bad image of agriculture if they took it from your lecture that *even* the homeless are able to do farmwork. I fear that the image further accelerates the alienation of people from agriculture.

This comment was written by a student from a farming family, as indicated elsewhere on the feedback sheet. The student appeared to be deeply attached to values learned from observing relatives' farmwork. It seemed the student was emphasizing that farming requires a specific skill set, a point on which farmers could take pride. Had the argument stopped there, I could have understood the comment as an objection to public disdain for the pride farmers take in their profession. This would align with the difficulty-conscious approach outlined in the previous two chapters.

However, the student appeared to have misinterpreted my words due to biases regarding homeless individuals. The first sentence is based on the assumption that homeless people have limited experience with "real society." More importantly, I never suggested that homeless laborers were performing farm tasks successfully to sustain the project. Instead, I highlighted the challenges they faced, even though some demonstrated remarkable resilience in coping with homelessness. In this regard, these homeless laborers were simply novice horticulturists, much like other urban citizens in similar circumstances. Thus, the word "even" in the second sentence carries an implied offense. If the student's concern about the image of agriculture is realized, the blame should fall not on homeless individuals but on those who discriminate against them. The entire citation ultimately expresses nothing but a negative view of a particular population.

This issue also raises a pedagogical concern. The growing practice of agriculture-welfare partnerships has begun to appear in educational settings (Gouda 2019). This trend introduces a new interdisciplinary academic responsibility, as agriculture and social welfare education have typically been taught separately. Educators now must address both misconceptions about agriculture held by urban dwellers and popular prejudices against socially excluded individuals. Effective teaching should emphasize both the demanding nature of farming and the significant potential of farm participants, despite the inevitable trade-offs. The next section will discuss how I navigated these challenges, effectively summarizing the key points of the preceding chapters.

7.2 Resolution of the Issue and Summary of Previous Chapters

In the years following the comment referenced earlier, I made it a practice to question students as outlined below. The inclusion of the term "disabled" was due to the fact that many farming activities provided support for job seekers with disabilities under the guise of partnership. This was also the case in my action research project, where several participants had developmental disabilities, as mentioned in Chap. 2. Moreover, social and medical workers became increasingly aware of the relationship between homelessness and mental disabilities, following two field surveys that confirmed the existence of this issue in Japan, as previously reported in Western countries (cf. Nishio et al. 2015).

> Concerning job creation in the agricultural sector, some people hold such a negative opinion: if someone imagines agriculture to be possible even for homeless or disabled people, they disparage farming skills disdaining professional farmers. State your views on this opinion for feedback.

The majority of recent feedback included one or both of the following points: (a) support for the negative opinion, reflecting farmers' professional pride in consistency with the earlier comment, and (b) denunciation of the opinion for its discriminatory stance toward homeless and disabled individuals. Generally, these arguments either upheld one sector's perspective from a utilitarian viewpoint, sacrificing the counterpart, or supported both sectors, leading to self-contradiction. The most frequent arguments, in order of prevalence, are as follows:

1. The understaffed agricultural sector cannot afford to be selective about its workforce.
2. Participants require instruction in knowledge and techniques.
3. Farm directors can assign simple tasks to unskilled participants.
4. Farmwork offers effective assistance to those excluded from the mainstream labor market, even if they remain unskilled.

The first argument appears as a counter to the previous comment. From a social inclusion standpoint, understandably, industries should not be selective about their workforce. However, this argument is based on a different premise: only sectors facing labor shortages must compromise. This creates an impression that individuals specializing in social welfare, who often come from non-rural backgrounds, undervalue agriculture. It is also conceivable that labor shortages may require even greater selectivity in human resources due to the urgent need for reorganization. As I discussed in Chap. 1, the partnership rests on a fragile theoretical foundation if the agricultural sector is considered suitable for collaboration with social welfare solely because of its labor shortage. It constitutes a demand to delegate the responsibility of integrating marginalized workers to an already understaffed sector. This could spark further discussion with students if the course design allowed for an ongoing relationship.

The difficulty-conscious approach is unrelated to supplementary recruitment for labor shortages. Chapter 2 highlighted how farming activities instilled a sense of liberation in participants. It also pointed to the communication between directors and participants, a critical analysis of which appeared in Chap. 3. This analysis revealed that the instructor-participant relationship could serve as a temporary substitute for essential skills needed for crop cultivation. However, I was uncertain about how to train participants to "take directions from crops" on their own. In this context, the second-most common argument makes sense but warrants no further discussion. Observing other practitioners of the agriculture-welfare partnership, as featured in Chap. 4, revealed alternative perspectives. I found myself compelled to explain why I pursued a path that aimed to enhance the autonomy of novice horticulturists. I feared that if farmwork tasks remained limited to simple duties, in line with the third-ranked argument, participants would miss out on the essential joys of growing crops. In response, Chap. 5 introduced concepts like the "operation-ladenness of observation" and the subsequent "I-did-it experience" to explain why I attribute the joy of horticulture to grappling with demanding tasks aimed at skill acquisition. Chapter 6 demonstrated that the difficulty-conscious approach could dispel potential misgivings of future participants, who would ultimately experience the satisfaction of farming, sometimes at public expense.

The fourth-ranked argument, however, seems to advocate for the opposite, an ease-driven approach, suggesting that it suffices for practitioners to inspire participants' sense of liberation even if they have not outgrown their instruction-taking position. This flexible approach does not contradict the difficulty-conscious model. A point of compromise between these two approaches emerges in the following student feedback: "Learners should respect professional farmers when learning farming skills." Rather than focusing on success or failure, the critical element may be developing an awareness of the inherent challenges of crop cultivation. An episode of the *teikei* organization mentioned in Case B4 of Chap. 4 well demonstrates the appropriateness of this comment. It may seem counterintuitive for retailers to sell vegetable seedlings in addition to vegetables per se, as vegetables may sell worse once customers have learned how to grow on their own. However, such experiences actually increase their appreciation for the skill of professional farmers. As customers attempt to grow the seedlings, they begin to understand the expertise involved in cultivation, thereby deepening their respect for the producers.

Another student offered support for a compromise between the two approaches, stating:

> Farming is the best-suited occupation for those who are excluded from the mainstream labor market because it is easier to start on an individual basis than other professions. If produce has turned out unmarketable due to their substandard growing or merchandizing skills, they can consume it for themselves to develop a self-sufficient lifestyle.

This comment adds a practical dimension to the previous ideological argument. Since beginners are unlikely to succeed in growing vegetables, they must recognize the disparity in skill between themselves and professional farmers, as some consumers in the *teikei* movement did in the aforementioned example. They could then seek modest guidance from professionals, marking a new phase in the agriculture-welfare partnership. This cooperation would enable them to become self-sufficient, thereby protecting themselves from the ongoing trend of primitive accumulation, as discussed in Chap. 6.

However, the first task for educators should be to encourage students to critically examine their perception of the issue. Returning to Chap. 1, it should be emphasized that the majority of society, rather than those facing exclusion, are responsible for the lack of respect toward farmers. A review of students' comments from this perspective reveals that none addressed the duties owed by the majority. Most students, aspiring to become members of the mainstream, risk focusing solely on moral judgment directed at the agricultural or social welfare sectors. These comments, regardless of their stance, are typically written in the third person, from the perspective of external observers. Educators should encourage students in introspection to recognize that they, too, are members of a society where the mainstream has marginalized both sectors.

This case is part of liberal education in universities, but the lessons here also apply to non-formal adult education or "social education" in Japan (cf. Ministry of Education, Culture, Sports Science and Technology 2010). As discussed in later sections of Chap. 6, the agriculture-welfare partnership could extend its influence beyond the two sectors if it directly addressed potential criticisms regarding public resources being allocated to farming activities for marginalized workers. The latest chapter examined several social work theories that could align with the partnership's practices, illustrating that the distinction between social work and education becomes less significant when practitioners seek to create a more supportive social environment.

7.3 Matters for Further Discussions

The key questions raised in the introductory chapter were whether the agriculture-welfare partnership can be integrated into social welfare systems and why the social welfare sector should collaborate with the agricultural counterpart. In essence, the inquiry revolves around what this partnership offers as an alternative. Literature has largely avoided providing a direct answer, perhaps due to the difficulty in rejecting familiar paradigms. However, the brief answer—summarizing the findings of subsequent chapters—is that the agriculture-welfare partnership can address the dualism between welfare and work, a persistent issue within social welfare policy. The Welfare-to-Work slogan in Japan reflects this dualism, wherein the aesthetic value of farmwork has been disregarded. This research challenges that perspective. The involvement of agriculture in social welfare can foster a fundamental transformation, helping to rectify the imbalanced urban-rural relationship.

The discussion of university students' feedback in my lectures serves to disclose one potential area where advocates may encounter misconceptions among the public. As noted in Chap. 1, I argued that the relationship between agriculture and welfare has primarily been ideological, with the sectors complementing each other only under unique economic circumstances, such as labor shortages in agriculture. This opportunity, seized by the Ministry of Agriculture, Forestry, and Fisheries (MAFF), has shaped the government's endorsement of sector collaboration with the Ministry of Health, Labor, and Welfare (MHLW). In the classroom, I simply presented this position. However, some students, drawn to the apparent logic of this reasoning, raised feedback suggesting that the agricultural sector's predicament—labor shortages coupled with unemployment—creates a mutually beneficial, non-political partnership. This view likely appeals because it seems pragmatic and non-partisan, at least within the Japanese context. I, too, have used similar logic to secure funding for the action research project.

Building on the theory of justification from a different perspective, the previous chapters offer grounds for further discussion. Agriculture is often described as multifunctional, creating external economies or by-products of material production (e.g., Van Huylenbroeck et al. 2007). The agriculture-welfare partnership, along with and care/social farming, highlights this multifunctionality distinctively (e.g., Zasada 2011; Nowack et al. 2022; Aizaki et al. 2024). A systematic review by Obeng et al. (2023) suggests that care farming and horticultural activities ultimately strengthen local economies by producing goods and services while caring for nature. An intrinsic quality of agriculture contributes to the evolution of social work practices, particularly in relation to the care of nonhuman life forms. Conversely, the social welfare sector could be the last resort to finally play a crucial role in embracing this external function.

Future discussions may address issues in social work that currently have no direct connection to farming. For instance, Molyneaux et al. (2011) argue that the labels "carer" and "cared-for" unnecessarily divide individuals who could otherwise collaborate. Horticulture for well-being offers a more practical solution than the alternative expressions proposed by these authors. Previous chapters explained that when a social worker integrates horticultural activities into their practice, clients assume the role of "carers" of nonhuman life forms, creating an egalitarian dynamic with the social worker, who also assumes the role of "carer."

Further, Parsell et al. (2017) contend that while contemporary social work emphasizes human agency, it remains distant from the neoliberal agenda, which seeks to gloss over structural and systemic forces underlying seemingly individual issues. This argument parallels the justification for autonomy in horticultural activities explored in earlier chapters. Carers, while responsible individuals, should also listen to the "voices" silenced by the aggressive and influential figures in neoliberalism.

Finally, Webb (2021) expands the discussion on human agency, suggesting that social work practice should be a process of "togethering," where both practitioners and service users engage through mutual entanglements with both human and non-human entities. This perspective aligns with the passivism embedded in the I-did-it experience theory, which recognizes the inherent uncertainties of life.

Fundamentally, the aesthetic value of agriculture is a by-product that contributes to the overall production of end products in an interconnected, cyclical relationship.

It is premature to assume that this by-product unilaterally flows from internal to external factors; the current externality may simply be an epistemological phenomenon within academic discourse. I propose that professional farmers may also have benefited from this value, a notion agronomists have been reluctant to uncover for several reasons.

First, as early as the beginning of the twentieth century, there was an idea that agricultural labor in diversified farm management could bring spiritual fulfillment in itself, rather than merely serving as a means to an external end. This concept, known in Japanese as the "agriculture-first principle," was later abandoned by some advocates, who apostatized in favor of militarism during the onset of the Fifteen Years War (1931–1945) in the Asia-Pacific region (Iwasaki 1992). Postwar agronomists often conflated this principle with militarism, considering it taboo and, at times, labeling it "agrarian nationalism" (cf. Havens 2015; Nakajima 2021).

Second, agronomists tended to emphasize the hardships of agriculture over its joys, despite the inseparability of both, as discussed in earlier chapters on the difficulty-conscious approach. This preference was partly due to the lingering influence of "agrarian nationalism" over discussions of the joy of farming. Furthermore, agronomists also confronted the cold fact that peasants lagged behind mainstream society in economic growth, feeling a sense of humiliation. The pervasive belief that work must be laborious further inhibited the recognition of farming's joyful aspects. As a result, agronomy has historically focused on production sales as the primary benefit for farming households.

Until recent decades, educators assumed that only individuals from rural backgrounds would engage in farming, expecting urban dwellers to remain detached from the realities of farm life and view themselves merely as sympathetic taxpayers or consumers, as implied in Chap. 2. In short, the agricultural sector has only recently begun to recognize the need for a deeper understanding of the intrinsic motivations behind farming.

This context explains why mainstream agronomists historically externalized the spiritual function of agriculture. The issue now is whether agronomy will continue to maintain this particular stance toward the by-product in question. There is no doubt that the agricultural sector is now motivated to engage a broader range of individuals than previously anticipated. In partnership with relevant organizations, the sector could conceptually reinternalize what has been externalized—provided that the agriculture-welfare partnership aims at more than just replenishing the workforce. Cooperative farmers would likely be willing to share their valuable skill-acquisition experiences with newcomers from social welfare organizations, viewing them as potential insiders rather than permanent outsiders. In this case, the boundary between internal and external would blur, necessitating a reconsideration of the concept of "externality." At the same time, "true productivism," a doctrine I introduced in the prior chapter, may undergo rigorous scrutiny before being integrated into progressive social work theories.

It is worth noting that externalities inherently invite "free riders." In the context of agriculture, examples may include urban residents such as educators, scholars, social workers, and their clients. If they remain unaware of their perceptions and behaviors in rural areas, this issue may warrant adult agricultural education. During my action research project, I discovered a prevalent myth embedded in urbanism: urban outsiders often view the healing power of rural landscapes as a gift from nature, rather than appreciating the role of human activity. In reality, this pastoral asset represents an external economy that local farmers have worked to maintain—initially for production purposes and, after retirement, for hospitality. Farmers feel the need to entertain urban visitors to preserve productive environments. However, the apathy of urban visitors toward agricultural production may eventually discourage rural residents, undermining the therapeutic benefits of the locality. Additionally, urban beneficiaries' experience of "farmwork" is sometimes merely a simulation, carefully fabricated by professional farm managers. If these beneficiaries then develop a misconception that such simulated farmwork contributes to material production and supports earnest farmers, the "ride" remains unpaid. In this sense, urban citizens still require education regarding the external economies they benefit from.

This volume is part of the Urban Agriculture Book Series. While "urban agriculture" typically refers to agricultural practices conducted in urban areas, we can also consider a temporal dimension, expanding the term to encompass agricultural activities performed by individuals from urban backgrounds, regardless of the location. In this sense, the term "urban" shifts to indicate limited farming experience, suggesting that individuals from urban backgrounds may require support from experienced practitioners. Previous chapters have further clarified the nature of this support.

7.4 Brief Mention of Social Inclusion

While the partnership counterpart has addressed several social issues originating from other industrial sectors, the agricultural sector has begun to do so in an unprecedented manner. This prompts a critical question: Will other players in the national or global economy remain indifferent spectators to social inclusion practices, accumulating an "industrial reserve army" (cf. Rigakos and Ergul 2011) to discard, or will they attempt to model themselves after the advanced case? If the majority opt for the former, history will likely repeat itself—the agricultural and social welfare sectors will again become marginalized, with their partnership serving as an excuse for the continued exclusion of disadvantaged workers from the mainstream.

This question revisits the issue raised in the introductory chapter regarding the relationship between "social inclusion" and "activation." There may be causality between the two if the dualism between welfare and work remains valid, as I previously denied. However, the extent to which the MAFF-MHLW federation assesses the gap between them remains unclear, partly due to "social inclusion" as a technical term difficult for government officials and ordinary citizens in Japan to comprehend (Dahl 2018). Furthermore, my action research, which initially seemed superficially parallel but was ultimately distinct from "activation," challenges this dualism. Findings from earlier chapters support this view, highlighting how farming activities differ from other occupations in that the activator is primarily nonhuman life forms. Consequently, the partnership movement and care/social farming may disentangle "social inclusion" from the artificial "activation" approach, despite official schemes framing the relationship between the two as an ends-means connection. This assertion does not suggest that both ideas should be discarded.

Rather, the question should be who is to be activated and included. While it is natural to consider the needs of the unemployed when formulating job creation schemes, those who criticize the jobless should also reflect on what jobs they can create for the public good. This process will either satisfy them or reveal the unreasonable nature of their prior claims. Future initiatives should target mainstream individuals, including critics, in this regard. My project exemplifies how social inclusion can engage citizens, starting with a farming household with an idle parcel of land. Homeless participants in need of employment opportunities proved to be reliable collaborators for young individuals receiving job support. These instances clearly illustrate the potential for citizen participation in social inclusion, serving as examples to mainstream society.

7.5 Postscript

The results of the 2022 survey, directed at practitioners of the agriculture-welfare partnership, including social welfare service providers and agricultural management agents (Japan Fund 2023), are presented here. Regarding the positive effects of the partnership, 87.5% of 776 social welfare enterprises reported that farming activities led to favorable changes in their service users. For example, more than half of the respondents noted that their clients "gained physical strength, enabling them to work longer hours," "became more motivated to perform tasks," "showed increased frequency of cheerful facial expressions," "became more confident through experiencing success," and "stabilized their nerves," listed in order of frequency.

These findings are consistent with those discussed in Chap. 2. Despite the methodological limitations imposed on my approach from an action research project with a small sample size, the present study seems to have reached a valid conclusion. In addition, I tried to compensate for the lack of political discussions in existing literature on the intersectoral partnership. However, my proposal for collaboration between agronomy and progressive social work theories is still remaining on the drawing board. Future research should examine its appropriateness through both empirical and theoretical investigations.

References[1]

Aizaki H, Akahori H, Yamamoto Y (2024) Measuring public importance for agricultural and rural multifunctionality: an application of case 1 best-worst scaling. Paddy Water Environ 23:49–60. https://doi.org/10.1007/s10333-024-00998-2

Dahl N (2018) Social inclusion of senior citizens in Japan: an investigation into the 'community-based integrated care system'. Contemp Jpn 30(1):43–59

**Gouda M (2019) About the present conditions and problems of welfare education utilizing "the Collaboration between Agriculture and Social Welfare": from the initiative of Kijimadaira Village Social Welfare Council. Acad Bull Nagano Univ 41(1):99–107

Havens TRH (2015) Farm and nation in modern Japan: agrarian nationalism, 1870–1940. Princeton University Press

***Iwasaki M (1992) Historical significances of the agriculture-first principle in return-to-the-farm movements ["Kinōhonshugi" no rekishitekiimi]. Ann Soc History Social Thought 16:103–114

***Japan Fund (2023) Results of survey of the partnership between agriculture and social welfare. https://noufuku.jp/wprs/wp-content/uploads/2023/04/r4-survey.pdf. Accessed 21 Oct 2024

Ministry of Education, Culture, Sports Science and Technology (2010) Social education in Japan. https://www.mext.go.jp/a_menu/shougai/shakai/1285290.htm. Accessed 29 Oct 2024

Molyneaux V, Butchard S, Simpson J, Murray C (2011) Reconsidering the term 'carer': a critique of the universal adoption of the term 'carer'. Aging Society 31:422–437

***Nakajima K (2021) Search for the path toward "agriculture symbiotic with nature": organic farming, natural farming, the peasantry ["Shizen to Tomo ni Aru Nōgyō" he no Michi o Saguru: Yūkinuōgyō, Shizen-nōhō, Shōnōsē]. Tsukuba Shobo, Tokyo

Nishio A, Yamamoto M, Horita R, Sado T, Ueki H, Watanabe T, Uehara R, Shioiri T (2015) Prevalence of mental illness, cognitive disability, and their overlap among the homeless in Nagoya, Japan. PLoS One 10(9):e0138052

Nowack W, Schmid JC, Grethe H (2022) Social dimensions of multifunctional agriculture in Europe: towards an interdisciplinary framework. Int J Agric Sustain 20(5):758–773

Obeng JK, Kangas K, Stamm I, Tolvanen A (2023) Promoting sustainable well-being through nature-based interventions for young people in precarious situations: implications for social work. A systematic review. J Happiness Stud 24:2881–2911

Parsell C, Eggins E, Marston G (2017) Human agency and social work research: a systematic search and synthesis of social work literature. Br J Soc Work 47:238–255

Rigakos GS, Ergul A (2011) Policing the industrial reserve army: an international study. Crime Law Soc Change 56:329–371

Van Huylenbroeck G, Vandermeulen V, Mettepenningen E, Verspecht A (2007) Multifunctionality of agriculture: a review of definitions, evidence and instruments. Living Rev Landscape Res 1(3) http://www.livingreviews.org/lrlr-2007-3. Accessed 4 Nov 2024

Webb SA (2021) Why agential realism matters to social work. Br J Soc Work 51:2964–2981

Zasada I (2011) Multifunctional peri-urban agriculture: a review of societal demands and the provision of goods and services by farming. Land Use Policy 28:639–648

[1] ***: in Japanese, **: in Japanese with an English title, *: in Japanese with an English title and summary.

Index

H. Tsunashima, *Farming and Social Inclusion in Japan*, Urban Agriculture, https://doi.org/10.1007/978-3-032-01598-3

MIX
Papier aus verantwortungsvollen Quellen
Paper from responsible sources
FSC® C105338

If you have any concerns about our products,
you can contact us on
ProductSafety@springernature.com

In case Publisher is established outside the EU,
the EU authorized representative is:
Springer Nature Customer Service Center GmbH
Europaplatz 3, 69115 Heidelberg, Germany

Printed by Libri Plureos GmbH
in Hamburg, Germany